THE DYNAMICS OF SHORELINE WETLANDS AND SEDIMENTS OF NORTHERN LAKE VICTORIA

Promoter: Prof. dr. P. Denny
                             Professor of Aquatic Ecology
                             UNESCO-IHE/Wageningen University
                             The Netherlands

Co-Promoters: Dr. J. van de Koppel
                             Scientific researcher mathematical modelling of
                             ecosystems
                             Centre for Estuarine and Marine Ecology
                             Netherlands Institute of Ecology, The Netherlands

                             Prof. dr. Frank Kansiime
                             Director
                             Makerere University Institute of Environment and
                             Natural Resources, Uganda

Awarding Committee: Prof. dr. C. Rossi
                             University of Siena, Italy

                             Prof. dr. J.T.A. Verhoeven
                             Utrecht University, The Netherlands

                             Prof. dr. M. Scheffer
                             Wageningen University, The Netherlands

                             Prof. dr. S. Uhlenbrook
                             UNESCO-IHE, The Netherlands

The dynamics of shoreline wetlands and sediments of northern Lake Victoria

DISSERTATION

Submitted in fulfilment of the requirements of
the Academic Board of Wageningen University and
the Academic Board of the UNESCO-IHE Institute for Water Education
for the Degree of DOCTOR
to be defended in public
on Wednesday, 20 December 2006 at 16:30 hours
in Delft, The Netherlands

by

NICHOLAS AZZA

born in Moyo, Uganda

CRC Press
Taylor & Francis Group
Boca Raton London New York

CRC Press is an imprint of the
Taylor & Francis Group, an **informa** business

CRC Press
Taylor & Francis Group
6000 Broken Sound Parkway NW, Suite 300
Boca Raton, FL 33487-2742

© 2006, Nicholas G.T. Azza
CRC Press is an imprint of Taylor & Francis Group, an Informa business

No claim to original U.S. Government works

ISBN-13: 978-0-4154-2649-7 (pbk)
ISBN-13: 978-1-1384-7489-5 (hbk)

Visit the Taylor & Francis Web site at
http://www.taylorandfrancis.com

and the CRC Press Web site at
http://www.crcpress.com

To the memory of Samuel Otuma (RIP)
A hydrological technician who drowned in Lake Victoria on 2 February 2002
while assisting in the collection of data for this dissertation.

Acknowledgements

Many people and organisations contributed greatly to the successful completion of this study. Although it is not possible to mention them all, their assistance, nonetheless, is most sincerely appreciated. Many thanks go to my Promoter Professor Patrick Denny, Co-promoter Dr. Johan van de Koppel and Local Supervisor Professor Frank Kansiime for their expert guidance and continuous encouragement.

My employer, the Ministry of Water, Lands and Environment in Uganda, through the Director of Water Development and Commissioner of Water Resources Management, provided me with access to office, laboratory, field research and transport facilities. I am most grateful for this facilitation. Special thanks go to Dr. F.L. Orach-Meza, John Wambede, Robert Kilama and the rest of the staff of the LVEMP Secretariat in Uganda who worked hard to ensure that funds were availed to me for fieldwork.

Office colleagues at the Water Resources Management Department in Entebbe assisted me in mobilising logistics for fieldwork, accompanied me during field excursions and occasionally lent a helping hand in sample analysis. I particularly wish to extend heartfelt thanks to the following: Saverino Mbarebaki, Simon Etimu, Abudallah Matovu, Livingstone Mabusi, Peter Obubu, Louise Mugisha, Richard Kyambadde, Damalie Ntwatwa, Richard Kabundama, Grace Ekuram, Sarah Majugu, Lydia Ebinu, Sarah Nabossa, Nelson Ssegawa, Juliet Lagua, James Erugudo (RIP), Siraje Musisi, Bart Baryakurahi and Janet Naguti (RIP).

I thank the Assistant Commissioner of Fisheries in Uganda, Mr. Francis Xavier Kiiza, for availing his boats and veteran coxswains Moses Kapere, Paul Obot, Chrisestom Kintu and Ramanzani Kawere. Without experienced hands spinning the nautical wheel and minding the boat engines, I would most probably have perished in the perilous waters of the great Lake Victoria. The support and cooperation of the Officers and Men of Uganda Peoples Defence Forces (UPDF) from Entebbe Marine Base is no less appreciated. I am indebted to Captain James Ssebulime, Sergeant Polycarp Setumba and Private Patrick Teise for facilitating and accompanying me during cross-lake cruises.

Lastly, I wish to acknowledge the unwavering moral support of my wife Rebecca, sons Ludger and Willibrord, and brother Alex. I must include in this category my dear friends Medard Hilhorst and Tini Brugge, as well as the family of Isaac and Alice Azabo, who have looked after me during my stay in The Netherlands. May Almighty God reward you for your kindness.

Contents

Contents

Chapter 1

Introduction

Introduction

Structure of the thesis

This thesis outlines the results of a doctoral study conducted on Lake Victoria. The report commences with an introduction (this chapter) that presents general background information on Lake Victoria and introduces the objectives of, and motivations for, the research. In chapters that follow (Chapters 2-6), the research carried out and its pertinent findings are described in detail. In Chapter 7, selected outcomes from Chapters 2 – 6 are reflected upon while in the final chapter (Chapter 8), conclusions and recommendations are outlined.

The Chapters 2 - 6 in which detailed findings are reported have been written in a self-contained manner: each contains, among others, the necessary background information needed to define the specific problem being addressed and set the work in perspective. This was done with a view to making the document reader-friendly: the reader is saved from having to read preceding chapters before moving to a chapter of interest, or continuously flipping back-and-forth to find cross-referenced information. Inevitably, however, this style introduces repetition, especially with regard to description of the study area and significance of the study. Notwithstanding, and considering that the repetitions are only a minor part of the chapters, it is felt that this style offers the most effective way of communicating the research findings to the wider scientific readership.

General background on Lake Victoria

Geography, origin and key hydrological features
Lake Victoria, which in past decades was known as the Victoria Nyanza (Worthington, 1930), is the largest lake in Africa, and the second largest freshwater lake in the world. The lake occupies an area of 68,870 km^2 under the territorial jurisdictions of Tanzania (49% of lake area), Uganda (45%) and Kenya (9%). Only Lake Superior (83,300 km^2) in North America has a larger surface area (Beeton, 1984). The lake has a correspondingly large catchment area (194,300 km^2) that extends to parts of Rwanda and Burundi (Figure 1.1).

The basin is perched high up (1,134 m above sea level) on the African craton and occupies a tectonic sag between the western and eastern arms of the African rift valley (Beadle, 1981). The basin of Lake Victoria is relatively flat and shallow with a maximum depth of 84 m and a mean depth of 40 m. From north to south it measures 400 km, and from east to west about 210 km, with approximately 3500 km of shoreline (Crul, 1998). Much of the shoreline is irregular. Shoreline development, which is the ratio of length of shoreline to the circumference of a circle with an area equivalent to that of the lake, is 3.7. Shallow bays and gulfs are numerous, particularly on the northern and southern shores. The lake also has archipelagos (i.e. clusters of islands), many located close to shore.

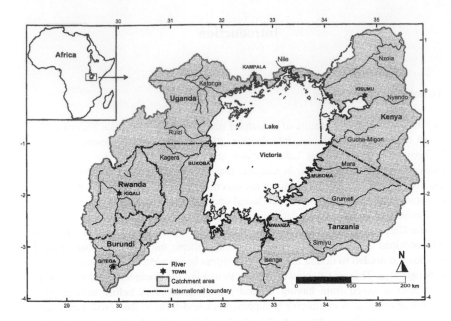

Figure 1.1: Map of Lake Victoria, its drainage basin and main rivers. Also shown are major towns in the catchment. Map coordinates are in decimal degrees.

Along many parts of the shoreline are large marginal wetlands with well developed flora dominated by the emergent macrophytes *Cyperus papyrus* L., *Miscanthidium violaceum* (Robyns) Schum., *Phragmites mauritianus* Kunth. and *Loudetia phragmitoides* (Peter) C.E. Hubbard. (Lind and Morrison, 1974). Within the catchment, especially in the northwestern part, wide boggy valleys completely filled with emergent herbaceous vegetation are also common (Beadle, 1981; Denny, 1993). In both the lakeshore and valley swamps, emergent vegetation has two growth forms: bottom rooted and floating-mat forms (Howard-Williams and Gaudet, 1985; Azza *et al.*, 2000).

Lake Victoria is of recent origin, having been formed about 750,000 years ago by earth movements starting in the late Miocene. Tectonic activity cut across the previous east-west drainage and tilted the western side rivers (Kagera and Katonga) reversing their flows eastwards into the area now occupied by the lake (Beadle, 1981). The most significant tectonic event, the late Pleistocene uprising of the Virunga volcanoes, created a divide which captured heavy rainfall from the highlands of Rwanda and Burundi and deflected it, via the Kagera River, into Lake Victoria. This water previously flowed westwards into Lake Edward. Earth movements, by giving rise to drowned valleys and causing the reversal and 'ponding back' of major rivers, also created conditions suitable for establishment of large papyrus swamps thereby producing a unique floral mosaic in the northwestern parts of the catchment (Beadle, 1981). The basin's history has been characterized by progressive uplift and tilting that has continued to the present day and produced its drowned morphology and distinctly irregular shoreline (Temple 1964, 1969; Bishop and Trendall, 1967; Johnson *et al.*, 2000). The epicentre of the lake is still shifting eastwards. Recent seismic reflection profiles (Johnson *et al.*, 1996; 2000) suggest the lake went through a number of desiccation events in its history.

Lake Victoria is the main reservoir maintaining the basic flow of the Nile, which is the only out-flowing river. The discharge of the Nile at the point of outflow, estimated at 31 km^3 yr^{-1} (Sene and Plinston, 1994), contributes 14% of the combined flows of the White and Blue Niles. The lake receives inflow from 24 tributaries, the largest of which is the Kagera River. The combined water input of the rivers amounts to less than 20% of the water entering the lake, the rest being provided by rainfall (Yin and Nicholson, 1998). Generally, direct rainfall and evaporation from the surface of the lake dominate its water budget. Water balance models of the lake [there are over 20 publications from the early work of Hurst and Phillips (1933) to the recent update by Tata *et al.*, (2004)] estimate groundwater inflow and outflow at zero. The lake has a flushing time of 140 years and residence time of 23 years (Crul, 1998). The main hydrological features of the basin have been summarized in Table 1.1.

Table 1.1. Main morphometric and hydrological parameters of Lake Victoria[1]

Parameter	Value (or names)
Elevation (a.m.s.l.)	1,134 m
Lake surface area	68,870 km^2
Volume	2,760 km^3
Mean depth	40 m
Maximum depth	84 m
Maximum length	400 km
Maximum width[2]	212 km
Mean width	174 km
Shoreline	3,440 km
Flushing time	140 yrs
Refill time/residence time)	23 yrs
Annual lake level fluctuations	0.01 – 1.66[3] m
Outflowing rivers	Nile
Major inflowing rivers:	
Tanzania	Mori, Mara, Grumeti, Simiyu, Isanga
Kenya	Nzoia, Sio, Yala, Kibos, Nyando,
	Sondu, Mongusi, Kuja, Awach, Kaboun
Uganda	Kagera, Katonga, Ruizi

[1] Source: Crul (1998) except where indicated otherwise
[2] Source: Spigel and Coulter (1996)
[3] Source: this study

Thermal stratification and water movements
Thermal stratification and lake hydrodynamics are of great limnological significance as they regulate lake productivity via their influence on the exchange of dissolved gases, dissolved minerals and particulate organic and inorganic matter. The annual cycle of thermal stratification in Lake Victoria shows considerable differences between nearshore and offshore regions (Worthington, 1930; Fish, 1957; Ochumba, 1983) and between different offshore regions (Graham, 1929, Newell, 1960;

LVEMP, 2002). In nearshore waters, stratification occurs mostly in the sheltered bays and inlets on the northern shore, where it is weak, strongly regulated by the diel cycle of solar heating, and is typically short-lived; breaking down completely during nocturnal mixing (Worthington, 1930; Fish, 1957; Ochumba, 1983).

In far offshore waters, stratification is very pronounced and occurs in three phases with between-year variation in the start and end of the cycle. The period from September to December is normally characterised by isothermal or very weakly stratified conditions in the open lake. This is followed by a warming phase from January to April during which multiple thermal discontinuities appear in the water column as the primary metalimnion is progressively displaced downwards. The cooling phase commences at the onset of the strong southeast trade winds in late May to early June and ends with the annual overturn of the lake and return to isothermal conditions (Fish, 1957; Talling, 1957, 1966; Newell, 1960).

During the period of strong offshore stratification, the water column is triply stratified into an epilimnion, metalimnion and hypolimnion, with a temperature range of 23.5 to 27 °C between hypolimnion and epilimnion. The density difference corresponding to this temperature range is comparable to a range between 6 and 13 °C in temperate lakes (Talling, 1957; Denny, 1972). The waters on the sheltered northern shore are warmer during the warming phase, and cooler during the cooling phase than offshore waters. Conversely, waters on the exposed western shore are cooler than offshore waters all year round (LVEMP, 2002). Waters on the western shore (located downwind of southeasterly winds) mix several times in a year.

Currents and water movements in the lake arise mostly from stress exerted by diurnal land/lake breezes, seasonal air currents and storms on the lake. Records of wind strength and direction at meteorological stations round the lake show a residual south component over the year, especially in the season May to October. Drift bottles released by Graham (1929) in this season revealed a lake-wide clockwise gyre. The reverse, an anticlockwise circulation, has also been shown during periods of predominantly easterly winds (Song et al., 2004).

Surface and internal seiches of a unimodal type are frequently induced in the lake by the cessation of strong winds (Bargman, 1953; Fish, 1957). Sustained southerly winds in the period of strong stratification cause a gradual northward displacement of surface water, a downwind tilting of the primary thermocline and a southward flow and upwelling of cold hypolimnetic waters (Fish, 1957; Newell, 1960; Talling, 1966). Cyclonic upwelling of cold bottom waters in the central parts of the lake has also been observed to accompany strong storms (Kitaka, 1972). Near the islands on the northern peripheries of the lake, differential heating and cooling of the shallow sheltered bays and adjacent deep offshore waters occasionally gives rise to profile-bound density currents (MacIntyre and Melack, 1995).

The economic and scientific significance of Lake Victoria

Lake Victoria and its surrounds are ecological systems of great value to man and nature. Human communities in the catchment are heavily dependent on the diversity of natural resources within the lake and catchment for food, water and way of life. The lake moderates regional climate, maintains the basic flow of the White Nile, serves as a reservoir for hydropower generation and provides riparian countries with a source of drinking water, cheap animal protein and means of transport. The lake's catchment is estimate to provide for the livelihood of about 33% of the combined populations of the three East African countries and about the same proportion (33%) of their combined domestic product (World Bank, 1996). Over 30 million people live in the catchment making it one of the most densely populated regions in Africa. Human population growth in the lake region is estimated at 3.5% p.a. - one of the highest in the world (Cohen *et al.*, 1996). A number of large towns, namely Musoma, Mwanza, Bukoba, Entebbe, Kampala, Jinja and Kisumu are located close to the shores of the lake. The lake supports incomes of the people in the catchment within a range of US $ 90-270 per capita per annum (World Bank, 1996).

Agriculture is the mainstay of the region's economy engaging over 80% of the population (Cohen *et al.*, 1996). There is a high degree of subsistence farming, with cash crops (coffee, tea, sugarcane, flowers and sisal) as well as food crops (plantains, pulses, cereals and tubers) being grown on mostly small farms. In Kenya and Uganda, the agricultural produce from the catchment is a significant proportion of national agricultural exports. The high productivity in the catchment is sustained by high rainfall and fairly fertile soils.

Lake Victoria is home to a diverse and unique assemblage of aquatic biota. Of special importance are the fish of the lake, of which there are over 500 species. The greater majority of the fish are endemic haplochromine cichlids (Greenwood, 1974; 1981). The native fish fauna were mostly living, between two million and ten thousand years ago, in the west-flowing rivers that later ponded back to form the lake. The explosive radiation of the cichlids into hundreds of species from a common ancestry in roughly 14,000 years (Johnson *et al.*, 1996) is the best example of evolutionary species radiation known to man (Goldschmidt and Witte, 1992).

The fishery of Lake Victoria is one of the most productive inland fisheries in the world (Witte *et al.*, 1995; FAO, 2004). Fish exports constitute a major source of revenue to the three East African countries. Once a multispecies fishery, the present fishery is dominated by three species: Nile tilapia (*Oreochromis niloticus*), a cyprinid (*Rastraneobola argentae*) locally known as *Mukene, Dagaa* or *Omena*, and the Nile perch (*Lates niloticus*) (Coulter *et al.*, 1986; Ogutu-Ohwayo, 1990, 1992; Witte *et al.*, 1999). The Nile perch and Nile tilapia were introduced from Lake Albert to boost commercial fisheries. Commercial catches peaked around 500,000 metric tons in 1990 (Bugenyi and Balirwa, 1998). The fishery, like the rest of the agriculture within the catchment, has a strong element of subsistence harvesting. Other economic activities in the catchment include livestock rearing, forestry and tourism.

Ecological changes in Lake Victoria
Ecological transformations with far-reaching implications have taken place in Lake Victoria in recent decades. These changes, which have been deduced from comparison of the modern condition of the lake with its historical status recorded in the works of Worthington (1930), Fish (1957) and Talling (1966), sum up as: warmer temperatures and more stable thermal stratification (Hecky, 1993; Lehman, 1998); increased hypolimnetic anoxia affecting nearly half of the lake's bottom and persisting for extended periods (Hecky, 1993; Hecky *et al.*, 1994; Bugenyi and Magumba, 1996), an increase in the incidence of catastrophic fish kills (Ochumba, 1987; Ochumba and Kibara, 1989; Ochumba, 1990); a twofold increase in pelagic

primary productivity and threefold increase in chlorophyll-a concentrations (Lehman and Branstrator, 1993; Mugidde, 1993); a tenfold decrease in soluble reactive silica concentrations and two- to fivefold increase in the concentrations of other plant nutrients (Hecky and Bugenyi, 1992; Hecky, 1993; Gophen *et al.*, 1993; Bugenyi and Magumba, 1996; Verschuren *et al.*, 1998); a replacement of diatoms as the dominant phytoplankton group by blue-green algae (Hecky, 1993; Mugidde, 1993; Balirwa, 1998); and a sharp rise in the population of Nile perch coinciding with a dramatic decline in other fish stocks and modification of pelagic food webs (Ogutu-Ohwayo, 1990, 1992; Ogutu-Ohwayo and Hecky, 1991; Goldschmidt and Witte, 1992; Kaufman, 1992; Witte *et al.*, 1992a,b; Witte *et al.*, 1995; Cohen *et al.*, 1996; Witte *et al.*, 1999).

Factors causing ecological change
While descriptions in literature on the ecological changes in Lake Victoria commonly go to great detail, amongst scientists, there is fluctuant conviction, and often disagreement, over the causative factors of change. Three competing hypotheses have been tendered to explain the changes: (1) trophic alterations caused by a cascade of predator-prey interactions triggered by introduction of the predatory Nile perch eliminated endemic herbivores and permitted the unrestrained growth of nuisance algae; (Witte *et al.*, 1992b; Gophen *et al.*, 1993; Goldschmidt *et al.*, 1993); (2) progressive anthropogenic disturbance of the lake's catchment has increased nutrient inflows to the lake and spurred the emergence of cyanobacteria dominance and expansion of areas affected by acute hypolimnetic anoxia (Ochumba, 1990; Bootsma and Hecky, 1993; Hecky, 1993; Muggide, 1993; Hecky *et al.*, 1994; Bugenyi and Magumba, 1996); and (3) a warming trend in the climate of East Africa accompanied by a decrease in the duration of strong winds has altered the lake's water column structure and mixing patterns, and triggered changes in the phytoplankton community structure (Hecky and Kling, 1987; Talling, 1987; Ochumba and Kibaara, 1989; Lehman, 1998).

Over time, consensus has formed amongst scientists on eutrophication and climate change being the leading causes of the ecological transformations in Lake Victoria (Lehman, 1998; Johnson *et al.*, 2000; Verschuren *et al.*, 2002). The case for cascading trophic interactions has been significantly weakened by careful examination of fish capture statistics (Ogutu-Ohwayo, 1990) and by recent paleolimnological evidence (Hecky, 1993; Lipiatou *et al.*, 1996) that together have demonstrated that the changes in lake biota preceded Nile perch introduction. Thus, following the logic that effect cannot precede cause, the predator could not have caused the ecosystem changes in the lake. Added to this, examination of trophic relations amongst organisms in the lake did not find clear evidence of the suggested simplification in food webs (Wanink and Witte, 1998; Wanink *et al.*, 2002). Moreover, further paloelimnological work established strong chronological linkage between land use change in the catchment and limnological transformations in the lake (Johnson *et al.*, 2000; Verschuren *et al.* 2002). Thus this view has effectively been discarded (Lehman, 1998). Notwithstanding, it is widely accepted that the voracious Nile perch is responsible for the extinction of several hundred species of endemic cichlid fishes.

Eutrophication is the most widely cited cause of the problems of Lake Victoria. The export of nutrients from the catchment is believed to have increased tremendously, from as early as the 1920s, through man's clearing of savannah and virgin forests, burning of bushes, practicing intensive agriculture and animal husbandry, and occupying lake shorelines for access to fish, (Hamilton, 1984; Cohen *et al.*, 1996; Crul, 1998). Direct verification of this hypothesis has been rendered difficult by the absence of historical measurements of nutrient exports from the catchment. However, Verschuren *et al.*, (2002) circumvent this problem by using human population size as a proxy indicator of anthropogenic soil and forest/savannah disturbance and its effect on nutrient fluxes. They showed (Figure 3 in their paper) that the timing and progress of lacustrine productivity increase matched human population growth and inferred agricultural and nutrient export activities within the catchment.

A burgeoning population growth is regarded as the underlying cause of the aquatic impacts from human activities in the catchment (Hecky, 1993). The explosion in human and livestock populations started around 1930 with the completion of the Uganda railroad, which stimulated plantation agriculture for the export of cash crops, and opened up the Lake Victoria region to settlement. Continued population growth through immigration and improved health conditions then started the pattern of large-scale deforestation and agricultural conversion that has continued to the present day. The commissioning of the hydropower station at Owen Falls Dam in 1954 further catalyzed economic growth and saw the emergence of new activities with a potential to further increase nutrient exports. These include the rapid expansion of towns and cities, increasing road construction, discharge of untreated municipal and industrial effluents and encroachment on wetlands (Hecky, 1993; Cohen *et al.*, 1996; Bugenyi and Balirwa, 1998; Crul, 1998; Kairu, 2001). The intensity of the new pressures is rising, being driven by the continuing rapid population rise and struggle of riparians for greater economic prosperity (World Bank, 1996).

The export of nutrients from the catchment may have been aggravated by the sudden rise in lake levels in the early 1960s that followed unusually heavy rains (Flohn, 1987; Sene and Plinston, 1994). The rise in lake levels caused extensive flooding of the shoreline, drowning of shoreline swamps and possible release of plant nutrients from flooded soils and drowned decomposing plant biomass.

Eutrophication had seemed most plausible and sufficient as an explanation because the ecosystem changes of 1960-1990 were widely viewed as being unique (Hecky, 1993). Recent reconstructions of past conditions in the lake, however, have indicated that modern events are not unique. This discovery raised the potential of climate change being a co-driver of the change process. Climate variation is linked to lake condition through mechanisms of heat budget and mixing regime, which are the consequences of atmosphere-lake interactions. Detailed fossil stratigraphy of diatoms during the last 10,000 years indicated that dominance of heavily silicified diatoms, proxy indicators of deep and sustained lake mixing conditions, rose and fell repeatedly and episodically over time periods of centuries or less (Stager *et al.*, 1997). Elemental and isotopic analysis of sedimentary organic matter also suggested that the Late Pleistocene to Holocene history of the lake was characterised by alternating periods of deep mixing and relative water column stability (Talbot and

Lærdal, 2000). In a recent effort to set up a hydrodynamic model for Lake Victoria (World Bank, 1999), it was realized that the lake's temperature structure and mixing regime were extremely sensitive to variations in meteorological forcing factors. Climate change, particularly change producing lower wind speeds and higher humidity, accordingly, has a strong potential to increase stratification and cause major alterations in the chemical and biological condition of the lake.

The Lake Victoria Environmental Management Project (LVEMP)
In recent years, the countries that share the lake have come together and, with assistance of donors, formulated programmes to halt the decline of the lake. The LVEMP is one such effort aimed at addressing the numerous threats to Lake Victoria. Financed by the Global Environmental Facility (GEF), the International Development Assistance (IDA) and the regional governments to a tune of US dollars 77.0 million, the LVEMP is the first phase of a long-term action plan to rehabilitate the lake's ecosystem. The objectives of the project are to: (1) maximize sustainable benefits to riparian communities through utilizing resources in the basin to generate food, employment and income, supply safe water, and sustain a disease-free environment; (2) conserve biodiversity and genetic resources; and (3) harmonize national environmental management efforts in order to achieve, to the extent possible, the reversal of increasing environmental degradation (World Bank, 1996). The project is expected to make its greatest economic impact in heading off development instability and the possible total collapse of the lake's valuable fisheries. The first phase of the project ended in 2003 and a second phase is in the offing.

Project activities, which were launched in 1996, were designed at two scales: pilot zone and lake-wide. The pilot zone activities focused on addressing environmental threats identified in selected parts of Lake Victoria and its catchment. In Uganda, the pilot zones were Sango Bay, Sese Islands, Murchison Bay, Napoleon Gulf and Berkeley Bay. Project activities included wetland conservation, reduction of sediment and nutrient (especially phosphorus) loading and reduction of domestic, municipal and industrial wastewater discharges to the lake.

The lake-wide level of activities have been more crosscutting and included actions such as the identification, quantification and monitoring of pollution sources in the catchment, modeling of lake circulation, harmonization of legislation and strengthening of institutional capacity. Through the last action - institutional capacity building - the project rehabilitated office and laboratory facilities and sponsored the training of a number of scientists in the region. The Promovendus is a beneficiary of the capacity building programme, and the research reported herein has been carried out within the framework of the LVEMP Project.

The two areas of interest

This study investigated two components of Lake Victoria's ecosystem: its shoreline wetlands and surficial sediments (the uppermost, usually unconsolidated, and easily erodible layer of sediment). Before discussion of study objectives, the function of the selected components and their influence on lake ecosystem dynamics will be considered in some detail.

The shoreline wetlands of Lake Victoria

The shoreline wetlands of Lake Victoria were studied but of what value are they? The shoreline wetlands are important ecosystems that have numerous functions and render valuable services to society (Costanza *et al.*, 1997). One of their key functions is the sustenance of a high diversity of living organisms (Denny, 1991, 1994). Generally, lakes that have some vegetation have a higher biodiversity than lakes that have no vegetation (Scheffer, 1998). Not only is a high diversity of living things good for conservation but it also leads to greater socio-economic benefits. It has been argued that a high biodiversity enhances the functioning of wetlands and boosts the services they render to society (Engelhardt and Ritchie, 2001). Thus maintaining biodiversity can be an effective way of sustaining societal benefits from wetlands.

Shoreline wetlands are colonised by biotic communities consisting of varying proportions of terrestrial and aquatic organisms, as well as organisms unique to the zone (Denny, 1985a, b, 1991; Wetzel, 1990). Macrophytes are among the most important structural elements in the shoreline communities. In the wetlands of Lake Victoria, two of the dominant emergent plants *Cyperus papyrus* and *Miscanthidium violaceum* form large floating rafts on which other plants anchor. Sedges and grasses are the most abundant plants associated with papyrus and *Miscanthidium*. There are also shrubs and small trees; climbing, creeping and scrambling herbs; ferns; and mosses in the swamps. At the lakeward fringes of the floating mats is usually a community of euhydrophytes dominated by water lilies (*Nymphaea*), water chestnuts (*Trapa*), water shields (*Brasenia*) water fringe (*Nymphoides*), duck lettuce (*Ottelia*), hornworts (*Ceratophyllum*), bladderworts (*Utricularia*), eelgrass (*Vallisneria*) and pondweed (*Potamogeton*). Lind and Morrison (1974) identified up to 63 species of plants occurring in papyrus swamps, many of them rooted in the mat. *Miscanthidium* swamps also have a large number of plant species though the floral diversity is somewhat less than that in papyrus swamps.

As well as plants, animals both vertebrate and invertebrate inhabit the shorelineal wetlands. Denny (1985b, 1991) argues that the biological and chemical interactions of the wetland ecotone create some of the most fertile habitats in the aquatic environment in terms of secondary production. The shoots of littoral macrophytes, Denny maintains, provide vast areas for colonisation by attached algae, bacteria and protozoa. The abundance of these organisms and the maintenance of oxygenated conditions in the zone by euhydrophytes (a collective term for submerged, floating-leaved and bottom-rooted aquatic macrophytes; Denny, 1985c), provides an ideal environment for many grazers particularly invertebrates and juvenile fish.

One of the better-known functions of the wetlands is their role as fish feeding, spawning and nursery grounds (Craig, 1987; Stephenson, 1990; Jude and Pappas, 1992; Kaufman and Ochumba, 1993; Brazne and Beals, 1997; Mnaya and Wolanski, 2002). This function, as suggested above, is largely dependent on an abundant supply of food of various types, which in turn is assured by the rich floral and faunal diversity in the littoral fringe, and the outflow of detritus from adjacent swamps (Wolf *et al.*, 1983; Denny, 1991; Matena, 1995; Simonian *et al.*, 1995; Laegdgaard and Craig, 2001). The variety of food available in shoreline wetlands is particularly important to juvenile fish that often change diet in the course of growth and development. At a global scale, a tight link has been suggested between nursery ground function of shoreline wetlands, and the size of adult fish populations (Turner, 1977). Thus the presence of shoreline vegetation is a critical requirement for sustaining the booming fisheries of Lake Victoria.

Closely connected with the food and nursery ground function is the use of shoreline wetlands by fauna as refuges from predation. Planktivorous zooplankton, and young and adult fish, hide in the dense foliage of euhydrophytes to escape capture by lurking predators (Burks et al., 2001; Laegdgaard and Craig, 2001). In some pelagic fauna, the strategy of predator avoidance leads to the striking phenomenon of diel horizontal migration (DHM) (Lauridsen and Buenk, 1996; Lauridsen et al., 1998; Burks et al., 2002), whereby organisms retreat to the shelter of littoral wetlands at dawn and stream back into open waters at dusk.

As well as offering structural refuge (i.e. protecting prey organisms through the structural complexity of the habitat) shoreline wetlands serve as physiological refuges. In Lake Nabugabo, a satellite lake within the Lake Victoria basin, many fish species escape Nile perch predation by hiding in the dark hypoxic interior of fringe floating swamps (Chapman et al., 1996; 2002, 2003). The darkness impairs visual predation while hypoxia keeps out predators unable to tolerate low oxygen conditions. It is probable that the shoreline wetlands of Lake Victoria function in a similar manner. The water flowing out from the interior of floating swamps commonly has low concentrations of dissolved oxygen. However, immediately beyond the swamps euhydrophytes reoxygenate the water column through their photosynthetic activity, creating an environment favourable for fauna unable to stand low oxygen conditions (Denny, 1985b; 1991).

A final function of shoreline wetlands is the regulation of fluxes of energy, water sediment and nutrients between the catchment and the open lake. Within the wetland, a number of physical, chemical and biological processes are in operation that help to purify inflowing water and buffer the lake from pollution (Kadlec and Tilton, 1979; Nichols, 1983; Wetzel, 1990; Kadlec and Knight, 1996; Jansson et al., 1998).

The shoreline wetlands, while helping to limit the inflow of materials from the catchment, themselves produce and export considerable quantities of detritus to adjacent nearshore waters. Tropical wetlands are among the most productive ecosystems in the world. Annual biomass production in papyrus swamps, for example, is estimated at 143 tons (dry matter) ha^{-1} (Thompson et al., 1979), which is greater than the productivity of many agricultural crops. Little of this production is consumed as living tissue. After the plants die, some of their remains accumulate as peat on the swamp floor but the larger proportion is exported to adjacent nearshore waters (Lind and Visser, 1962; Gaudet, 1976; Howard-Williams and Gaudet, 1985; Kansiime and Nalubega, 1999). These outflows of detritus are important in sustaining a detritus-based food web in the nearshore zone. Where favourable conditions for sediment accumulation exist, the detritus exports can also be important in biogeochemical cycling of major elements. However, the exports can have detrimental effects on nearshore water quality.

The above, from a socio-economic point of view, have been more or less positive functions of shoreline wetlands. On the negative side of things, dense growths of shoreline vegetation can interfere with recreation activities and hinder fishing activities by blocking fishermen's passages. Large islands of floating mat can break off and block ports thereby obstructing water transport. The wetlands can also harbour vectors of tropical diseases such as malaria, schistosomiasis and tripanosomiasis.

What is the present situation with regard to wetlands research in Africa? Research into the ecology of African wetlands commenced in the first quarter of the twentieth century under the encouragement of colonial governments. In 1985 a book, *The ecology and management of African wetland vegetation* (Denny, 1985d), was published that provided a valuable synthesis of the knowledge accumulated since the beginning of systematic research on African wetlands. The book reviewed the biology of the different groups of aquatic macrophytes, the structure and function of the ecosystems of which they are part, and the general conservation and management of the wetlands. Considerable space is devoted to examining the structure and function of shoreline systems including the floating swamps of Lake Victoria. The book has therefore been an important source of information for this study and is frequently referred to herein.

The amount of research since publication of the African wetlands book has decreased, but scientific interest in wetlands has continued (see for example Azza *et al.*, 2000; Gichuki *et al.*, 2001a,b; Masifwa *et al.*, 2001; Kipkemboi *et al.*, 2002; Kansiime *et al.*, 2003; Ssegawa *et al.*, 2004). Despite the considerable amount of ground covered in past decades, some basic ecological principles remain unexplored, such as the factors regulating the vertical and horizontal zonation of shoreline vegetation. Along the shoreline of Lake Victoria, the distribution of wetlands, and hence the distribution of beneficial services derivable from them, is not uniform. Some stretches of shore have large areas of wetlands while others are devoid of vegetation. What could be the reason for this? Additionally, Lake Victoria's wetlands have as a conspicuous feature the occurrence of floating root mats on which many plants anchor. How does the mat arise, and of what ecological significance is it? Cleary, a need for continued research exists.

Lake sediments

In the foregoing discussion, shoreline wetlands were shown to influence lake ecosystem dynamics in multiple ways. Sediments too, can profoundly affect the chemical and biological processes within a lake. This mainly results from their ability to bind and transport nutrients, heavy metals and other micropollutants (Hakanson and Jansson, 1983; Harper, 1992). In the Laurentian Great Lakes, the resuspension of sediments has been reported to occasionally introduce fluxes of nutrients to the water column that are much greater than the sum of fluxes from all other sources (Eadie and Robbins, 1987). Such events can result in massive blooms of algae, catastrophic fish kills from upwelling of oxygen-depleted hypolimnetic water, lake-wide dispersal of sediment-bound pollutants and water treatment problems at water supply works (Harper, 1992).

What is the state of knowledge with respect to Lake Victoria? Very little can be said about the sediments of Lake Victoria mainly because, unlike the wetlands, they have not been intensively investigated. A number of studies that looked at sediments were carried out under the IDEAL (International Decade for the East African Lakes) programme (Johnson, 1993). However, the objectives of IDEAL were to enhance understanding of the paleoclimatology, paleohydrology and paleolimnology of the East African lakes. Hence, the studies did not investigate sediments per se but used the sediment record to make inferences on other subjects of interest (i.e. the ancient climatology, hydrology and limnology of the lakes) (see for example Stager et al., 1986; Talbot and Livingstone, 1989; Beuning et al., 1997, 1998; Johnson et al., 1998, 2000; Talbot and Lærdal, 2000; Stager and Johnson, 2000; Nicholson and Yin, 2001; Verschuren et al., 2000, 2002).

Outside of IDEAL, not many studies have looked at sediments. The few that have done so have mainly focused on the mineralogy of the sediments (Mothersill, 1976) and their content of toxic heavy metals (Onyari and Wandiga, 1989; Makundi, 2001; Campbell et al., 2003; Kishe and Machiwa, 2003; Ramlal et al., 2003).

From the above, it is clear that knowledge on the sediments of Lake Victoria is still in its infancy. Among others, the basic physico-chemical characteristics of the sediments (which determine the mechanisms by which they are transported and their ability to bind nutrients and micropollutants; Hakanson and Jansson, 1983), their within-lake distribution, and the main processes driving their distribution, are not well known. Such knowledge is necessary for a better appreciation of the influence of sediments on the ecological dynamics of the lake.

It is possible that the sediments of the lake, like lacustrine sediments elsewhere, have a strong influence on the ecosystem dynamics. Studies on drainage patterns and geomorphodynamics of the lake basin have shown that the major western river, the Kagera, delivered enormous quantities of terrigenous sediments to Lake Victoria in ancient times (Temple, 1964; Doornkamp and Temple, 1966; Bishop and Trendall, 1967; Temple, 1969; Temple and Doornkamp, 1970). Present day estimates of sediment loading by the Kagera are around 728×10^3 tons yr^{-1} (Department of Water Resources Management, Uganda, unpublished data). Historical loading was probably much greater considering the extensive sand deposits on the western coastline of the lake.

The bulk of sediments delivered by the Kagera river were deposited around its delta and the northwesten coast. However, some of the sediment, especially fine-grained sediments, must have penetrated beyond the shoreline zone to accumulate in the lake's deep offshore basin. Holtzman and Lehman (1998) have argued that on account of apatite-bearing rocks (rich in phosphorous) being widely distributed in the region, sediment transport by the Kagera and other feeder rivers must have resulted in continuous fertilisation of the lake with respect to phosphorous. Other authors have further suggested that in the presence of abundant phosphorous, nitrogen became limiting and induced an emergence of cyanobacteria dominance over other phytoplankton during the Holocene (Johnson et al., 2000; Talbot and Lærdal, 2000), as well as in modern times (Talling, 1966; Hecky, 1993; Lehman and Brandstrator, 1993; Mugidde, 1993).

The interaction between the lake's hydrodynamic regime and bottoms sediments could be having important effects on the system. Strong mixing events affecting large parts of the lake have a potential to intermittently recharge the pelagic reservoir of phosphorous and perpetuate cyanobacteria dominance. The possibility of lake-wide sediment dispersal with associated effects on ecosystem dynamics have not been closely examined.

The present study

Problem statement

Lake Victoria is a vital resource under stress from mainly anthropogenic activities. A few regional initiatives have been launched, and more are planned, to help halt its decline and sustain its beneficial services to society and nature. For any lake restoration measure to be effective, it must be founded on a sound knowledge of the past and present state of the lake, the complex and dynamic interactions between its various biotic and abiotic components, and the key factors driving its decline. Many gaps exist in the knowledge on Lake Victoria making it difficult to arrive at the most critical interventions required at the different stages of restoration. Two of the areas where information is scanty are shoreline wetlands and lake sediments. Both wetlands and bottom sediments have a potential to exert a strong influence on the functioning of the lake ecosystem. Considerable knowledge on wetland ecology exist in the western world but differences between temperate and tropical regions makes inappropriate the direct transfer of generalisations. The tropics receive stronger solar radiation, have higher annual temperatures and continuous growing seasons, all of which may modify the influence of a given factor on community structure and function. With regard to lake sediments, present generalisations on distribution are derived mostly from observations in smaller basins and may not directly apply to a basin of Lake Victoria's size.

Objective

The objective of the research is to determine the way in which the shoreline wetlands and sediments of northern Lake Victoria are spatially distributed, the factors regulating their distribution and, their potential influence on lake ecosystem function.

Research questions

The main question, which follows from the above objective, is:

> *How are the shoreline wetlands and sediments of Lake Victoria spatially distributed, what factors control their spatial distribution and what is their influence on lake ecosystem function?*

Subordinate to this question, a set of key questions have been formulated in five areas as follows:

1. *Wave exposure and spatial distribution of costal wetlands*
   How are shoreline wetlands spatially distributed? Is their distribution controlled by wave exposure? Do the morphometric characteristics of a bay affect the extent to which it is colonized by shoreline vegetation? What factors control the development of floating mats of emergent vegetation in costal wetlands? What is the purpose and ecological significance of floating mats formed by shoreline emergent vegetation?

2. *Shoreline wetlands and fish species decline*
   Could the littoral wetlands of Lake Victoria have undergone progressive degradation as a result of eutrophication potentially contributing to the observed decline in fish species? What are the mechanisms by which eutrophication impairs the function of littoral wetlands as fish habitats?

3. *Sediment distribution and transport mechanisms*
   How are surficial sediments of northern Lake Victoria spatially distributed? Are generalizations on sediment distribution in small temperate lakes applicable to large lakes in general, and a great tropical lake located in a region of strong winds in particular? What are the main ways by which sediments within Lake Victoria are moved around and re-distributed?

4. *Causal factors of ecological change*
   How robust are the theories on the ecological changes in Lake Victoria?

5. *General*
   What are the ways in which the shoreline wetlands and surficial sediments of Lake Victoria influence its ecosystem dynamics?

Approach and organisation of the research
This research addresses a real world problem: the degradation of a great lake, Lake Victoria. It aims to facilitate the process of defining management actions that can reverse the decline of the lake, by filling knowledge gaps. However, in filling the knowledge gaps, it follows a scientific research approach, endeavouring to test the limits of established scientific theory. Although the study focuses on a single lake, Lake Victoria, attempts are made to present and discuss the findings of the study in a broader limnological context.

The research was broken down into five blocks, each forming a theme for a chapter. In Table 1.2, the focus of each block, specific problem addressed and specific objective behind the investigation is presented. In a wrap-up discussion (Chapter 7) inferences from key findings from the five blocks (Chapters 2 to 6) are reflected upon. The last chapter (Chapter 8) lists the conclusions and recommendations of the study.

Study area
The study was mostly, but not entirely, restricted to the northern half of Lake Victoria due to the logistical difficulties of covering the entire lake. Each chapter contains a map showing the locations where samples or measurements were taken. It is assumed that the studied portion of the lake gives a reasonable representation of general conditions in the whole lake.

Rationale
This research is necessary because Lake Victoria is a resource of great socio-economic and ecological significance. Bridging gaps in our knowledge of the ecological dynamics of the lake could help improve our ability to manage it and

reverse its decline. Many of Africa's other great lakes, such as Lakes Albert, Kivu, Tanganyika and Malawi, are located within the same or comparable climatic zones and are facing similar anthropogenic pressures. The generalizations from this study could therefore be applied in the management of the other African great lakes. Generally, much less is known about these other lakes than is known about Lake Victoria.

Originality and creativity

This dissertation makes attempts at originality and creativity in as far as:
  (a) It focuses on topics on Lake Victoria that have previously not been tackled.
  (b) It brings new evidence to bear on old issues.
  (c) It follows, in some parts, a testing-out approach seeking to identify limits of previously established generalizations.
  (d) It follows, in some parts, an exploratory approach attempting to find plausible reasons for observed patterns.
  (e) It is cross-disciplinary making use of a wide range of analytical methodologies.

Table 1.2. The five blocks representing the focal areas of research

Block	Specific problem statement	Specific objectives
Block 1		
Spatial distribution of shoreline wetlands: the influence of wave exposure and embayment characteristics (Chapter 2)	1. Quantitative aspects of wave exposure on distribution tropical shoreline vegetation are not clearly known.	1. To test the hypothesis that vegetation distribution along shores a tropical great lake (Lake Victoria) is controlled by wave exposure through: a. Examining the qualitative patterns of shoreline vegetation distribution and comparing wave exposure in vegetated and non-vegetated sites; b. Investigating the variability of shoreline swamp area along a gradient of wave exposure; c. Assessing how well wave exposure, acting in concert with shoreline morphometric characteristics, is able to predict the lakeward limit of shoreline vegetation advancement.
	2. Little consideration has been given to the possible influence of embayment characteristics on shoreline vegetation distribution.	2. To investigate the possible influence of bay characteristics on shoreline vegetation distribution.

Table 1.2 (Continued from previous page).

Chapter	Specific problem statement	Specific objectives

Block 2
 	
Spatial distribution of shoreline wetlands: the influence of floating mats
 	
(Chapter 3)
	1. A basic understanding is lacking of the significance of the floating mat growth form. Specifically:
	  a. It is not known why shoreline emergent vegetation sometimes grow in the bottom-rooted form, and at other times in the floating-mat form;
	  b. It is not clearly known what causes the change from bottom-rooted to floating-mat form, or the mechanisms by which the change occurs;
	  c. It is not known what benefits or limitations the floating-mat form confers upon shoreline emergent vegetation.
		1. To improve the basic understanding of the biology of emergent floating mats, particularly:
		  a. The factors controlling their development;
		  b. The mechanisms by which they are initiated;
		  c. Their function or purpose; and
		  d. Their possible influence on the shoreline distribution of emergent plants.

Block 3 Shoreline wetlands: possible role in fish species decline (Chapter 4)	1. The causal factors and mechanisms for the catastrophic decline in fish species in Lake Victoria has not been clearly identified. 2. The possibility that the decline was caused in part by littoral wetland degradation has not been thoroughly investigated	1. To improve knowledge on the causal factors for ecosystem change in Lake Victoria by examining the possibility that fish species decline was due in part to littoral wetland degradation. 2. To identify potential mechanisms by which the interaction between eutrophication and littoral wetlands could cause decline in fish species.

Block 4
 	
Surficial sediments: spatial distribution and transport mechanisms
 	
(Chapter 5)
	1. In an eutrophic system like Lake Victoria, the transport and distribution of sediments, which often have attached nutrients, has important consequences with regard to the dynamics of the entire lake system. However, the main processes and mechanism by which sediment is distributed in the lake are not known. Furthermore, the applicability of generalizations on lake sediment distribution has not been demonstrated in Lake Victoria.	1. To test the applicability to Lake Victoria of three generalizations on lake sediment distribution, namely that:
		  a. Sediment resuspension results almost entirely from the action of wind-induced waves;
		  b. Sediments are spatially distributed in a focused manner;
		  c. Surficial sediment characteristics vary systematically with increasing water depth and fetch.
		2. To determine the likely mechanism for sediment distribution in Lake Victoria.

Block 5 Causal factors for the ecological changes in Lake Victoria: evidence from sediment cores (Chapter 6)	1. Eutrophication and climate change have been hypothesized as being responsible for the recent ecological transformations in Lake Victoria. However, these theories have not been rigorously examined due to a paucity of monitoring data.	1. To improve understanding of the recent limnological changes in Lake Victoria and test aspects of the theories that present eutrophication and climate change as the causal factors of the ecological changes in the lake.

References

Azza, N.G.T., Kansiime, F., Nalubega, M., and Denny, P. (2000) Differential permeability of papyrus and *Miscanthidium* root mats in Nakivubo swamp, Uganda. *Aquatic Botany* 67(3): 169-178.

Balirwa J. S. Lake Victoria wetlands and the ecology of the Nile Tilapia, *Oreochromis niloticus*, Ph.D. thesis. 1998. The Netherlands, Agricultural University of Wageningen.

Bargman, D. J. Evidence of a natural oscillation of Lake Victoria and waves caused by meteorological phenomena. Annual Report 1953, No. 13. 1953. Uganda Department of Hydrological Survey.

Beadle, L.C. The inland waters of tropical Africa. 1981. London, Longman.

Beeton, A.M. (1984) The world's Great Lakes. *Journal of Great Lakes Research* 10(2): 106-113.

Beuning, K.R.M., Kelts, K., and Stager, J.C. (1998) Abrupt climatic changes associated with the Younger Dryas interval in Africa. In *Environmental Change and Response in East African Lakes*. Lehman, J.T. (ed). Dordrecht: Kluwer Academic Publishers, pp 147-156.

Beuning, K.R.M., Kelts, K., Ito, E., and Johnson, T.C. (1997) Paleohydrology of Lake Victoria, East Africa, inferred from $^{18}O/^{16}O$ ratios in sediment cellulose. *Geology*, 25: 1083-1086.

Bishop, W.W. and Trendall, A.F. (1967) Erosion-surfaces, tectonics and volcanic activity in Uganda. *Quart. J. Geol. Soc. London* 132: 238-252.

Bootsma, H.A. and Hecky, R.E. (1993) Conservation of the African Great Lakes: A limnological perspective. *Conservation Biology* 7(3): 644-656.

Brazner, J.C. and Beals, E.W. (1997) Patterns in fish assemblages from coastal wetland and beach habitats in Green Bay, Lake Michigan: a multivariate analysis of abiotic and biotic forcing factors. *Canadian Journal of Fisheries and Aquatic Sciences* 54: 1743-1761.

Bugenyi, F.W.B. and Balirwa, J.S. (1998) East African species introductions and wetland management: Sociopolitical dimensions. In *Science in Africa: Emerging water management issues*. Schoneboom, J. (ed). Philadelphia: American Association for the Advancement of Science.

Bugenyi, F.W.B. and Magumba, K.M. (1996) The present physico-chemical ecology of Lake Victoria, Uganda. In *The Limnology, Climatology and Paloeclimatology of the East African Lakes*. Johnson, T.C. and Odada, E.O. (eds). Amsterdam: Gordon and Breach, 141-154.

Burks, R.L., Jeppesen, E., and Lodge, D.M. (2001) Littoral zone structures as *Daphnia* refugia against fish predators. *Limnology and Oceanography* 42(6): 230-237.

Burks, R.L., Lodge, D.M., Jeppesen, E., and Lauridsen, T.L. (2002) Diel horizontal migration of zooplankton: costs and benefits of inhabiting the littoral. *Freshwater Biology* 47(3): 343-365.

Campbell, L.M., Hecky, R.E., Nyaundi, J., Mugidde, R., and Dixon, D.G. (2003) Distribution and foodweb transfer of Mercury in Napoleon and Winum Gulfs, Lake Victoria, East Africa. *Journal of Great Lakes Research* 29(2): 267-282.

Chapman, L.J., Chapman, C.A., and Chandler, M. (1996) Wetland ecotones and refugia for endangered fishes. *Biological Conservation* 78: 263-270.

Chapman, L.J., Chapman, C.A., Nordlie, F.G., and Rosenberger, A.E. (2002) Physiological refugia: swamps, hypoxia tolerance and maintenance of fish diversity in the Lake Victoria region. *Comparative Biochemistry and Physiology Part A* 133: 421-437.

Chapman, L.J., Chapman, C.A., Schofield, P.J., Olowo, J.P., Kaufman, L., Seehausen, O., and Ogutu-Ohwayo, R. (2003) Fish fauna resurgence in Lake Nabugabo, East Africa. *Conservation Biology* 17(2): 500-511.

Cohen , A.S., Kaufman, L., and Ogutu-Ohwayo, R. (1996) Anthropogenic threats, impacts and the conservation strategies in the African Great Lakes - A review. In *The Limnology, Climatology and Paloeclimatology of the East African Lakes*. Johnson, T.C. and Odada, E.O. (eds). Amsterdam: Gordon Breach, 575-624.

Costanza, R., d'Arge, R., de Groot, R., Faber, S., Grasso, M., Hannon, B., Limburg, K., Naeem, S., O'Neill, R.V., Paru-elo, J., Raskin, R.G., Sutton, P., and van der Belt, M. (1997) The value of the world's ecosytem services and natural capital. *Nature*, 387: 253-260.

Coulter, G.W., Allanson, B.R., Bruton, M.N., Greenwood, P.H., Hart, R.C., Jackson, P.B.N., and Ribbink, A.J. (1986) Unique qualities and special problems of African Great Lakes. *Env. Biol. Fish.* 17: 161-184.

Craig, J. F. The biology of perch and related fish. 1987. Portland, Oregon, Timber Press.

Crul, R.C.M. Management and conservation of the African Great Lakes. 1998. Paris, Studies and Reports in Hydrology No. 59. UNESCO Publishing.

Denny, P. (1994) Biodiversity and wetlands. *Wetlands Ecology and Management* 3(1): 55-61.

Denny, P. (1972) The significance of the pycnocline in tropical lakes. *African Journal of Tropical Hydrobiology and Fisheries*, 2: 85-89.

Denny, P. (1993) Eastern Africa. In: *Wetlands of the World. I*, D.F. Whigham, D. Dykyjová and S. Hejn (eds). Kluwer Academic Publishers, Dordrecht pp. 32-46.

Denny, P. (1991) African wetlands. In *Wetlands*. Finlayson, M. and Moser, M. (eds). Oxford: Facts on File Publishers, 115-148.

Denny, P. (1985a) Submerged and floating-leaved aquatic macrophytes (euhydrophytes). In *The ecology and management of African wetland vegetation*. Denny, P. (ed). Dordrecht: Dr. W. Junk Publishers, 19-42.

Denny, P. (1985b) The structure and functioning of African euhydrophyte communities: The floating-leaved and submerged vegetation. In *The ecology and management of African wetland vegetation*. Denny, P. (ed). Dordrecht: Dr. W. Junk Publishers, 125-152.

Denny, P. (1985c) Wetland vegetation and associated life forms. In *The ecology and management of African wetland vegetation*. Denny, P. (ed). Dordrecht: Dr. W. Junk Publishers, 1-18.

Denny, P. (1985d) *The Ecology and Management of African Wetland Vegetation*, Dr. W. Junk Publishers, Dordrecht 256 pp.

Doornkamp, J.C. and Temple, P.H. (1966) Surface, drainage and tectonic instability in part of southern Uganda. *the Geographical Journal* 132: 238-252.

Eadie, B.J. and Robbins, J.A. (1987) The role of particulate matter in the movement of contaminants in the Great Lakes. In *Sources and fates of aquatic pollutants*. Hites, R. and Eisenreich, S. (eds). Washington D. C.: American Chemical Society, 319-364.

Engelhardt, K.A.M. and Ritchie, M. (2001) Effects of macrophyte species richness on wetland ecosystem functioning and services. *Nature* 411: 687-689.

FAO (2004) *State of the World Fisheries and Aquaculture*. Rome: FAO.

Fish, G.R. (1957) A seiche movement and its effects on the hydrology of Lake Victoria. *Fish. Publ.* 10: 1-68.

Flohn, H. (1987) East African rains of 1961/62 and the abrupt change of the White Nile discharge. *Paleoecology of Africa* 18: 3-18.

Gaudet, J.J. (1976) Nutrient relationships in the detritus of a tropical swamp. *Archiv fuer Hydrobiologie* 78: 213-239.

Gichuki, J., Guebas, D.F., Mugo, J., Rabour, C.O., Triest, L., and Dehairs, F. (2001) Species inventory and the local use of the plants and fishes of the lower Sondu Miriu wetland of Lake Victoria, Kenya. *Hydrobiologia* 458(1-3): 99-106.

Gichuki, J., Triest, L., and Dehairs, F. (2001) The use of stable carbon isotopes as tracers of ecosystem functioning in contrasting wetland ecosystems of Lake Victoria, Kenya. *Hydrobiologia* 458(1-3): 91-97.

Goldschmidt, T., Witte, F., and Wanink, J. (1993) Cascading effects of the introduced Nile perch on the detritivorous/phytoplanktivorous species in the sublittoral areas of Lake Victoria. *Conservation Biology* 7(3): 686-700.

Goldschmidt, T. and Witte, F. (1992) Explosive speciation and adaptive radiation of haplochromine cichlids from Lake Victoria: An illustration of the scientific value of a lost species flock. *Mitteilungen-Internationale Vereininging fur Theoretische und Angewandle Limnologie* 23: 101-107.

Gophen, M., Ochumba, P.B.O., Pollingher, U., and Kaufman, L.S. (1993) Nile perch (*Lates niloticus*) invasion in Lake Victoria (East Africa). *Verh. Internat. Verein. Limnol.* 25: 856-859.

Graham, M. The Victoria Nyanza and its Fisheries. A report on the fishing surveys of Lake Victoria. 1929. London, Crown - Agents for the Colonies.

Greenwood, P. H. The Cichlid Fishes of Lake Victoria, East Africa. 1974. London, British Museum of Natural History.

Greenwood, P. H. The Haplochromine Fishes of the East African Lakes. Collected papers on their taxonomy, biology and evolution. 1981. München, Kraus International Publications.

Hakanson, L. and Jansson, M. (1983) *Principles of lake sedimentology*. Berlin: Springer-Verlag.

Hamilton, A.C. Deforestation in Uganda. 1984. Nairobi, Oxford University Press.

Harper, D.M. (1992) *Eutrophication of freshwaters: principles, problems and restoration*. New York: Chapman & Hall.

Hecky, R.E. and Bugenyi, F.W.B. (1992) Hydrology and chemistry of the African Great Lakes and water quality issues. *Mitt. Verein. Internat. Limnol.* (23): **23**, 45-54.

Hecky, R.E., Bugenyi, F.W.B., Ochumba, P.O.B., Talling, J.F., Mugidde, R., Gophen, M., and Kaufman, L. (1994) Deoxygenation of the deep water of Lake Victoria, East Africa. *Limnology and Oceanography* 39(6): 1476-1481.

Hecky, R.E. and Kling, H.J. (1987) Phytoplankton ecology of the great lakes in the rift valleys of Central Africa. *Arch. Hydrobiol. Beih.* 25: 197-228.

Hecky, R.E. (1993) The eutrophication of Lake Victoria. *Verh. Internat. Verein. Limnol.* 25: 39-48.

Holtzman, J. and Lehman, J.T. (1998) Role of apatite weathering in the eutrophication of Lake Victoria. In *Environmental Change and Response in East African Lakes*. Lehman, J.T. (ed). Dordrecht: Kluwer Academic Publishers, pp 89-98.

Howard-Williams, C. and Gaudet, J.J. (1985) The structure and functioning of African swamps. In *The ecology and management of African wetland vegetation*. Denny, P. (ed). Dordrecht: Dr. W. Junk Publishers, 153-176.

Hurst, H.E., and Phillips, P. (1933) The Nile Basin, Vol. III. Ten-day mean and monthly mean guage readings of the Nile and its tributaries. Government Press, Cairo.

Jansson, A., Folke, C., and Langaas, S. (1998) Quantifying the nitrogen retention capacity of natural wetlands in the large scale drainage basin of the Baltic Sea. *Landscape Ecology* 13: 249-262.

Johnson, T.C., Chan, Y., Beuning, K.R.M., Kelts, K., Ngobi, G., and Verschuren, D. (1998) Biogenic silica profiles in Holocene cores from Lake Victoria: implications for lake level history and initiations of the Victoria Nile. In *Environmental Change and Response in East African Lakes*. Lehman, J.T. (ed). Dordrecht: Kluwer Academic Publishers, pp 75-88.

Johnson, T.C., Scholz, C.A., Talbot, M.R., Kelts, K., Ricketts, R.D., Ngobi, G., Beuning, K.R.M., Ssemmanda, I., and McGill, J.W. (1996) Late Pleistocene desiccation of Lake Victoria and rapid evolution of cichlid fishes. *Science* 273: 1091-1093.

Johnson, Thomas C. IDEAL: An International Decade for the East African Lakes. Science and Implementation Plan. 1993. Bern, Switzerland, PAGES Core Project Office.

Johnson, T.C., Kelts, K., and Odada Eric (2000) The Holocene history of Lake Victoria. *Ambio* 29(1): 2-11.

Jude, D.J. and Pappas, J. (1992) Fish utilisation of Great Lakes coastal wetlands. *Journal of Great Lakes Research* 18(4): 651-672.

Kadlec, R.H. and Knight, R.L. (1996) *Treatment Wetlands*. New York: Lewis Publishers, CRC.

Kadlec, R.H. and Tilton, D.L. (1979) The use of freshwater wetlands as a tertiary wastewater treatment alternative. *C.R.C. Critical Reviews in Environmental Control* 9(2): 185-212.

Kairu, J.K. (2001) Wetland use and impact on Lake Victoria, Kenya region. *Lakes & Reservoirs: Research and Management* 6(2): 117-125.

Kansiime, F. and Nalubega, M. Waste Water Treatment by a natural wetland: The Nakivubo Swamp, Uganda: processes and implications. 1999. Balkema Publishers, Rotterdam, The Netherlands, Wageningen Agricultural University and International Institute of Hydraulic and Environmental Engineering, The Netherlands.

Kansiime, F., Nalubega, M., van Bruggen, J.J., and Denny, P. (2003) The effect of wastewater discharge on biomass production and nutrient content of *Cyperus papyrus* and *Miscanthidium violacium* in the Nakivubo wetland, Kampala, Uganda. *Water Science and Technology* 48(5): 233-240.

Kaufman, L. (1992) Catastrophic change in species-rich freshwater ecosystems. The lessons of Lake Victoria. *BioScience* 42(11): 846-858.

Kaufman, L. and Ochumba, P. (1993) Evolutionary and conservation biology of cichlid fishes as revealed by faunal remnants in northern Lake Victoria. *Conservation Biology* 7(3): 719-730.

Kipkemboi, J., Kansiime, F., and Denny, P. (2002) The response of *Cyperus papyrus* (L.) and *Miscanthidium violaceum* (K. Schum.) Robyns to eutrophication in the natural wetlands of Lake Victoria, Uganda. *African Journal of Aquatic Science* 27(1): 11-20.

Kishe, M.A. and Machiwa, J.F. (2003) Distribution of heavy metals in sediments of Mwanza Gulf of Lake Victoria, Tanzania. *Environment International* 28(7): 619-625.

Kitaka, G.E.B. (1972) An instance of cyclonic upwelling in the southern offshore waters of Lake Victoria. *African Journal of Tropical Hydrobiology and Fisheries*, 2: 85-92.

Laegdsgaard, P. and Craig, J. (2001) Why do juvenile fish utilise mangrove habitats? *Journal of Experimental Marine Biology and Ecology* 257(2): 229-253.

Lauridsen, T.L. and Buenk, I. (1996) Diel changes on the horizontal distribution of zooplankton in the littoral zone of two shallow eutrophic lakes. *Archiv fur Hydrobiologie* 137(2): 161-176.

Lauridsen, T.L., Jeppesen, E., Sondergaard, M., and Lodge, D.M. (1998) Horizontal migration of zooplankton: predator-mediated use of macrophyte habitat. In *Structuring Role of Submerged Macrophytes in Lakes*. Jeppesen, E., Sondergaard, M., Sondergard, M., and Kristoffersen, K. (eds). New York : Springer-Verlag, 233-239.

Lehman, J.T. and Brandstrator, D.K. (1993) Effects of nutrients and grazing on phytoplankton of Lake Victoria. *Verh. Internat. Verein. Limnol.* 25: 850-855.

Lehman, J.T. (1998) Role of climate in the modern condition of Lake Victoria. *Theor. Appl. Climatol.* 61: 29-37.

Lind, E.M. and Morrison, E.S. East African vegetation. 1974. London, Longman Group.

Lind, E.M. and Visser, S.A. (1962) A study of a swamp at the northern end of Lake Victoria. *Journal of Ecology* 50: 599-613.

Lipiatou, E., Hecky, R.E., Eisenreich, S.J., Lockhart, L., Muir, D., and Wilkinson, P. (1996) Recent ecosystem changes in Lake Victoria reflected in sedimentary natural and anthropogenic organic compounds. In *The Limnology, Climatology and Paleoclimatology of the East African Lakes.* Johnson, T.C. and Odada, E. (eds). Amsterdam: Gordon and Breach, 523-541.

LVEMP (Lake Victoria Environmental Management Project). The Lake Victoria Integrated Water Quality and Limnology Study. 2002. Dar-es-Salaam, Tanzania, LVEMP Regional Secretariat.

MacIntyre, S. and Melack, J.M. (1995) Vertical and horizontal transport in lakes: linking littoral, benthic and pelagic habitats. *Journal of the North American Benthological Society*, 14: 599-615.

Makundi, I.N. (2001) A study of heavy metal pollution in Lake Victoria sediments by energy dispersive x-ray fluorescence. *Journal of Environmental Science and Health* A36(6): 909-921.

Masifwa, W.F., Twongo, T., and Denny, P. (2001) The impact of water hyacinth *Eichhornia crassipes* (Mart) Solms on the abundance and diversity of aquatic macroinvertebrates along the shores of northern Lake Victoria, Uganda. *Hydrobiologia* 452(1-3): 79-88.

Matena, J. (1995) The role of ecotones as feeding grounds for fish fry in a Bohemian water supply reservoir. *Hydrobiologia* 303: 31-38.

Mnaya, B. and Wolanski, E. (2002) Water circulation and fish larvae recruitment in papyrus wetlands, Rubondo Island, Lake Victoria. *Wetlands Ecology and Management* : 133-143.

Mothersill, J.S. (1976) The mineralogy and geochemistry of teh sediemnts of northwestern Lake Victoria. *Sedimentology*, 23: 553-565.

Mugidde, R. (1993) The increase in phytoplankton primary productivity and biomass in Lake Victoria (Uganda). *Verh. Internat. Verein. Limnol.*, 25: 846-849.

Newell, B.S. (1960) The hydrology of Lake Victoria. *Hydrobiologia*, 15: 363-383.

Nichols, D.S. (1983) Capacity of natural wetlands to remove nutrients from wastewater. *Journal of Water Pollution Control Federation* 55(5): 495-505.

Nicholson, S.E. and Yin, X. (2001) Rainfall conditions in equatorial East Africa during the nineteenth century as inferred from the record of Lake Victoria. *Climate Change* 48: 387-398.

Ochumba, P.B.O. (1987) Periodic massive fish kills in the Kenyan part of Lake Victoria. *Water Quality Bull.* 12: 119-122.

Ochumba, P.B.O. (1990) Massive fish kills in the Nyanza gulf of Lake Victoria, Kenya. *Hydrobiologia* 208: 93-99.

Ochumba, P.B.O. and Kibaara, D.I. (1989) Observations on blue-green algal blooms in the open waters of Lake Victoria, Kenya. *Afr. Jl. Ecol.* 27: 23-24.

Ogutu-Ohwayo, R. (1992) The purpose, costs and benefits of fish introductions, with specific reference to the Great Lakes of Africa. *Mitt. Internat. Verein. Limnol.* 23: 33-44.

Ogutu-Ohwayo, R. and Hecky, R.E. (1991) Fish introductions in Africa and some of their implications. *Can. J. Fish. Aquat. Sci.* 48(Suppl. 1): 8-12.

Ogutu-Ohwayo, R. (1990) The decline of the native fishes of Lake Victoria and Kyoga (East Africa) and the impact of introduced species, especially the Nile perch, *Lates niloticus*, and the Nile tilapia, *Oreochromis niloticus*. *Environmental Biology of Fishes* 27: 81-86.

Onyari, J.M. and Wandiga, S.O. (1989) Distribution of Cr, Pb, Cd, Zn, Fe and Mn in Lake Victoria sediments, East Africa. *Bulletin of Environmental Contamination and Toxicology* 42: 807-813.

Ramlal, P.S., Bugenyi, F.W.B., Kling, G.W., Nriagu, J.O., Rudd, J.W.M., and Campbell, L.M. (2003) Mercury concentrations in water sediment and biota from Lake Victoria, East Africa. *Journal of Great Lakes Research* 29(2): 283-291.

Scheffer, Martin. Ecology of Shallow Lakes. 1998. London, Chapman & Hall.

Sene, K.J. and Plinston, D.T. (1994) A review and update of the hydrology of Lake Victoria in East Africa. *Hydrological Sciences Journal* 39(1): 47-63.

Simonian, A., Tatrai, I., Biro, P., Paulovits, G., G.-Toth, L., and Lakatos, G. (1995) Biomass of planktonic crustaceans and the food of young cyprinids in the littoral zone of Lake Balaton. *Hydrobiologia* 303: 39-48.

Song, Y., Semazzi, F.H.M., Xie, L. and Ogallo, L.J. (2004) A coupled regional climate model of the Lake Victoria basin of East Africa. *International Journal of Climatology*, 24: 57-75.

Spigel, R.H. and Coulter, G.W. (1996) Comparison of hydrology and physical limnology of the East African Great Lakes: Tanganyika, Malawi, Victoria, Kivu and Turkana (with reference to some North American Great Lakes). In *The Limnology, Climatology and Paloeclimatology of the East African Lakes*. Johnson, T.C. and Odada, E.O. (eds). Amsterdam: Gordon and Breach Publishers, 103-140.

Ssegawa, P., Kakudidi, E., and Muasya, M. (2004) Diversity and distribution of sedges on multivariate environmental gradients. *African Journal of Ecology* S1: 21-33.

Stager, J.C., Cumming, B., and Meecher, L. (1997) A high-resolution 11,400-year diatom record from Lake Victoria, East Africa. *Quaternary Research* 47: 81-89.

Stager, J.C. and Johnson, T.C. (2000) A 12,400 ^{14}C yr offshore diatom record from east central Lake Victoria, East Africa. *Journal of Paleolimnology* 23(4): 373-383.

Stager, J.C., Reinthal, P.N., and Livingstone, D.A. (1986) A 25,000-year history for Lake Victoria, East Africa, and some comments on its significance for the evolution of cichlid fishes. *Freshwater Biology* 16(1): 15-19.

Stephenson, T.D. (1990) Fish reproductive utilisation of coastal marshes of Lake Ontario near Toronto. *Journal of Great Lakes Research* 16: 71-81.

Talbot, M.R. and Livingstone, D.A. (1989) Hydrogen index and carbon isotopes of lacustrine organic matter as lake level indicators. *Palaeogeography, Palaeoclimatology, Palaeoecology* 70((1-3)): 121-137.

Talbot, M.R. and Lærdal, T. (2000) The Late Pleistocene-Holocene paleolimnology of Lake Victoria, East Africa, based on elemental and isotopic analyses of sedimentary organic matter. *Journal of Paleolimnology* 23: 141-164.

Talling, J.F. (1987) The phytoplankton of Lake Victoria (East Africa). In *Phycology of large lakes of the world*. Munawar, M. (ed). *Arch. Hydrobiol. Beih. Ergebn. Limnol.*, 25:229-256.

Talling, J.F. (1966) The annual cycle of stratification and phytoplankton growth in Lake Victoria (East Africa). *Int. Revue ges. Hydrobiol.* 51(4): 545-621.

Talling, J.F. (1957) Some observations on the stratification of Lake Victoria. *Limnology and Oceanography*, 2: 213-221.

Tata, E., Sutcliffe, J., Conway, D. and Farquharson, F. (2004) Water Balance of Lake Victoria: update to 2000 and climate change modelling to 2100. *Hydrological Sciences Journal*, 49: 563-574.

Temple, P.H. (1964) Evidence of lake-level changes from the northern shoreline of Lake Victoria, Uganda. In *Geographers and the Tropics: Liverpool Essays*. Steel, R.W. and Prothero, R.M. (eds). London: 31-56.

Temple, P.H. Raised strandlines and shorelines evolution in the area of Lake Nabugabo, Masaka District, Uganda. Wright, H. E. Proc. INQUA 16. 1701, 119-129. 1969. National Academy of Sciences.

Temple, P.H. and Doornkamp, J.C. (1970) Influences controlling lacustrine overlap along the north-western margins of Lake Victoria. *Zeitschrift für Geomorphologie* 14(3): 301-317.

Thompson, K., Shewry, P.R., and Woolhouse, H.W. (1979) Papyrus swamp development in the Upemba Basin, Zaire: studies of population structure in *Cyperus papyrus* stands. *Botanical Journal of the Linnaean Society* 78: 299-316.

Turner, R.E. (1977) Intertidal vegetation and commercial yields of penaerial shrimp. *Transactions of the American Fisheries Society* 106(5): 411-416.

Verschuren, D., Edgington, D.N., Kling, H.J., and Johnson, T.C. (1998) Silica depletion in Lake Victoria: Sedimentary signals at offshore stations. *Journal of Great Lakes Research* 24(1): 118-130.

Verschuren, D., Johnson, T.C., Kling, H.J., Edgington, D.N., Leavit, P.R., Brown, E.T., Talbot, M.R., and Hecky, R.E. (2002) History and timing of human impact on Lake Victoria, East Africa. *Proc. R. Soc. Lond. B* 269: 289-294.

Verschuren, D., Laird, K., and Cumming, B.F. (2000) Rainfall and drought in equatorial East Africa during the past 1,100 years. *Nature* 403: 410-414.

Wanink, J.H. and Witte, F. Intraguild predation in Lake Victoria: was the food web really simplified after Nile perch boom? P. 254. 1998. African Fishes and Fisheries Diversity and Utilisation. Poissons et Peches Africains Diversite et Utilisation.

Wanink, J.H., Katunzi, E.F.B., Goudswaard, K.P.C., Witte, F., and Van Densen, W.L.T. (2002) The shift to smaller zooplankton in Lake Victoria cannot be attributed to the 'sardine' Rastrineobola argentea (Cyprinidae). *Aquatic Living Resources* 15: 37-43.

Wetzel, R.G. (1990) Land-water interfaces: metabolic and limnological regulators. *Verh. Internat. Verein. Limnol.* 24: 6-24.

Witte, F., Goldschimidt, T., Goudswaard, P.C., Ligtvoet, M.J.P., van Oijen, M.J.P., and Wanink, J.H. (1992a) Species extinction and concomitant ecological changes in Lake Victoria. *Netherlands Journal of Zoology* 42: 214-232.

Witte, F., Goldschmidt, T., and Wanink, J.H. (1995) Dynamics of the haplochromine cichlid fauna and other ecological changes in the Mwanza gulf of Lake Victoria. In *The impact of species changes in African Lakes*. Pitcher, T. and Hart. P. J. B. (eds). London: Chapman and Hall, 83-110.

Witte, F., Goldschmidt, T., Wanink, J., van Oijen, M., Goudswaard, K., Witte-Maas, E., and Bouton, N. (1992b) The destruction of an endemic species flock: quantitative data on the decline of the haplochromine cichlids of Lake Victoria. *Environmental Biology of Fisheries* 34: 1-28.

Witte, F., Goudswaard, P.C., Katunzi, E.F.B., Mkumbo, O.C., Seehausen, O., and Wanink, J.H. (1999) Lake Victoria's ecological changes and their relationships with the riparian societies. In *Ancient lakes: Their cultural and biological diversity*. Kawanabe, H., Coulter, G.W., and Roosevelt, A.C. (eds). Belgium: Kenobi Productions, 189-202.

Wolf, E.G., Morson, B., and Fucik, K.W. (1983) Preliminary studies of food habits of juvenile fish, China Poot Marsh and Potter Marsh, Alaska, 1978. *Estuaries* 6(2): 102-114.

World Bank. Preparation of a preliminary Lake Victoria physical processes and water quality model: Final report by the Consultant. 1999. Entebbe, The World Bank.

World Bank. Project Document for the Lake Victoria Environmental Management Project. 1996. Washington, DC, Global Environmental Facility (GEF), The World Bank.

Worthington, E.B. (1930) Observations on the temperature, hydrogen-ion concentration, and other physical conditions of the Victoria and Albert Nyanzas. *Int. Rev. ges. Hydrobiol. Hydrogr.* 24: 328-357.

Yin, X. and Nicholson, S.E. (1998) The water balance of Lake Victoria. *Hydrological Sciences Journal* 43(5): 789-811.

Shoreline vegetation distribution in relation to wave exposure and bay characteristics in Lake Victoria

Submission to a scientific journal based on this chapter:

Nicholas Azza, Johan van de Koppel, Patrick Denny and Frank Kansiime. Shoreline vegetation distribution in relation to wave exposure and bay characteristics in a tropical great lake, Lake Victoria, *Journal of Tropical Ecology,* submitted on October 14, 2006.

Shoreline vegetation distribution in relation to wave exposure and bay characteristics in Lake Victoria

Abstract

The concept that wind-wave exposure is a major regulator of vegetation distribution within lakes was investigated. Along a 675 km stretch of shore in northern Lake Victoria (Uganda), the pattern of vegetation distribution in relation to shoreline features, and the variation of shoreline swamp area along a wave exposure gradient were examined. The ability of wave exposure, when combined with bay morphometric characteristics, to predict the lakeward limit of vegetation distribution was assessed. Data was collected through a shoreline survey and from maps. Maximum effective fetch, computed from topographic maps, was used as a surrogate for wave exposure. The results amplify the hypothesis that wave exposure is an important regulator of the within-lake distribution of vegetation. Shoreline plants were found to either occupy stretches of shore shielded by coastal islands or hidden by convolutions of the lake margin. Swamp area declined exponentially with increasing wave exposure and appeared to be affected by the magnitude as well as the frequency of wave disturbance. Of the coastal characteristics examined, bay area had the strongest influence on the lakeward expansion of vegetation. Wave exposure, acting together with bay area and bay aspect ratio, accounted for 65.1% of the variance in the limit of lakeward vegetation advancement.

Keywords: vegetation distribution, wave exposure, bay characteristics, disturbance, Lake Victoria

Introduction

Shoreline plants have a major influence on the ecosystem functioning of water bodies they fringe (Carpenter, 1983). Their distribution in lakes is regulated by numerous environmental factors and by competition from neighbours (Pearsall, 1920; Hutchinson, 1975; Spence, 1982; Keddy, 2000). Wave exposure is one of the environmental factors with a strong influence on shoreline vegetation distribution. It affects vegetation directly by uncovering seeds, uprooting seedlings and damaging mature plants, and indirectly by producing coarser, nutrient deficient substrates, or burying established plants (Keddy, 1982; Foote and Kadlec, 1988; Coops et al., 1991; Kennedy and Bruno, 2000; Riis and Hawes, 2003). Through the above effects, wave exposure produces distinct patterning in shoreline vegetation, with upwind shores having better-developed flora than downwind shores (Pearsall, 1920; Hutchinson, 1975; Spence, 1982). Wave exposure also limits the lakeward advance of shoreline vegetation (Keddy, 1983).

Present knowledge on the influence of wave exposure on shoreline plant distribution is limited in some areas. The quantitative impact of wave exposure on tropical shoreline vegetation is one area where knowledge is limited. Current knowledge of the influence of wave exposure is derived mainly from temperate studies, which in many cases is applied unquestionably to tropical lakes (Denny, 1985b). Some disparity, nevertheless, must be expected due to differences between the two regions. In the inner tropics, for example, plant recruitment and growth occur all year at near-optimal rates. This may allow faster recovery of plants damaged during seasonal wind perturbations so that in the long-term, the influence of wind and waves is less clearly manifest. Early studies on the influence of wave exposure on the littoral vegetation of tropical lakes, such as Denny (1973), were largely qualitative although it was recognised that the dynamics of exposed and sheltered shores supported different vegetation pattern. Little consideration has also been given to the possible influence of embayment characteristics on vegetation distribution. Recent studies that have examined the effect of coastal morphometric features on vegetation distribution have mainly addressed water depth and/or littoral slope (see for example Duarte and Kalff, 1986; Chambers, 1987; Coops *et al.*, 1994; Hudon *et al.*, 2000; Riis and Hawes, 2003). Considering that the shape of a water body affects the way in which waves are generated (CERC, 1984; Maas et al, 1997), bays of certain shapes, sizes or aspect ratios may be expected to be better than others at moderating or amplifying the impacts of wind and waves and hence could have greater or lesser influence on vegetation distribution.

In this study the distribution of vegetation along the shores of a tropical great lake in relation to wind and wave exposure was investigated. The study (a) examined the qualitative pattern of shoreline vegetation distribution and compared wave exposure in vegetated and non-vegetated parts of the shore; (b) investigated the variation of shoreline vegetation along a gradient of wave exposure; and (c) assessed how well wave exposure, acting in concert with coastal morphometric characteristics, was able to predict the lakeward limit of vegetation advancement.

Materials and Methods

Study area

Lake Victoria, found in East Africa, is the second largest freshwater lake in the world. It has a saucer-shaped basin with a surface area of 68,870 km^2, maximum depth of 84 m and mean depth of 40 m (Crul, 1998). Its margins are convoluted giving rise to numerous shallow bays fringed in many parts by large tropical swamps. A swamp according to Howard-Williams and Gaudet (1985) is a boggy, seasonally or permanently flooded area with dense growths of tall herbaceous hydrophytes. Swamp vegetation may be bottom-rooted or anchored in floating root mats (Denny, 1993).

The swamps of Lake Victoria are dominated by the sedge *Cyperus papyrus* L. and grass *Miscanthidium violaceum* (K. Schum.) Robyns which grow in association with shrubs and trees; climbing, creeping and scrambling herbs; ferns; and mosses. At the lakeward boundary of the swamps, a zone of euhydrophytes (a collective word for submerged, floating-leaved and bottom-rooted aquatic macrophytes; Denny, 1985c) is often found, dominated by *Nymphaea* (Tourn.) L., *Trapa* L., *Potamogeton* L., *Ceratophyllum* L. and *Vallisneria* Mich ex L. Surface-floating plants including *Pistia* L. and *Eichhornia* Kunth may fringe the emergents and, if in profusion, shade out the euhydrophytes. Some vegetated margins have no fringe swamps but feature other marginal plants, mainly the reed *Phragmites mauritianus* Kunth., the shrub *Sesbania sesban* L. and the rhizomatous, low-growing perennial grass *Panicum subalbidum* Kunth. More detailed descriptions of the shoreline vegetation can be obtained from Lind and Morrison (1974) and Denny (1985a). Lake Victoria is fresh, alkaline and eutrophic (LVEMP, 2002).

The study was conducted in the Ugandan part of the lake on a 675 km stretch of shore between Goma and Berkeley bays (Figure 2.1). This stretch is positioned downwind of prevailing southerly winds and has ragged as well as even margins, sheltered and exposed shores, a sprinkling of coastal archipelagos and fringing swamps of varying sizes.

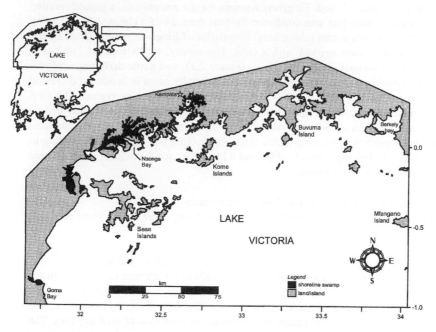

Figure 2.1. Map of the studied shore of Lake Victoria showing the distribution of shoreline swamps. The swamps inhabit shores behind coastal islands or nest in the crooks of bay indentations. Shoreline wetland cover was extracted from FAO Africover maps prepared from remote sensing images taken between 1999 and 2001.

Determination of exposure to wind-induced wave action

Maximum effective fetch was used as the surrogate of wave exposure and was calculated from 1:50,000 topographical map sheets (series Y732 of Uganda) using the formula (Håkanson and Jansson, 1983; CERC, 1994)

$$E_f = \frac{\sum x_i \cos a_i}{\sum \cos a_i} \tag{2.1}$$

where E_f (km) is the effective fetch (i.e. the open water distance over which wind-induced waves build), x_i (km) is the fetch length or straight-line distance from the point of fetch measurement to land or an island, and a_i (degrees) is the angle from the wind direction azimuth in 6° increments from +42° to -42°. Maximum effective fetch is the outcome of equation (2.1) when the wind direction azimuth is oriented in such a way as to give the longest possible straight-line distance from a measurement point to land or island.

Vegetation distribution along the shore

To determine the qualitative pattern of shoreline vegetation distribution, a map of the shoreline was inspected and the location of vegetated areas noted in relation to coastal morphometric features and position of coastal islands. Wave exposure in vegetated and non-vegetated parts of shore was also measured and compared with the Mann-Whitney U test. To generate points for the comparison, a ground-truthing survey of the shoreline was conducted by boat from 24-28 February 2003. During the survey, the horizontal (alongshore) boundaries of fringe vegetation (floating and bottom-rooted) were marked with a GPS. The survey, which covered 190 km of shore between Goma and Nsonga bays (Figure 2.1), was particularly important for locating non-swamp shoreline vegetation that normally occur in bands too narrow to display on the available 1:50,000 topographic maps of the lake region. The GPS coordinates were used to mark on topographic map sheets the start and end of vegetation patches encountered on the shore. In the stretches of vegetated and non-vegetated patches, ticks were added from which fetch was computed according to equation 2.1. Spacing intervals of 0.5 and 5 km were used for ticks in vegetated and bare-shore areas respectively.

Variation of swamp area along a gradient of wave exposure
The studied shore has a myriad of bays: some large, some small, some with and some without shoreline swamps. Seventy-eight (78) of the largest swamp-fringed bays were selected for study. At three points across the mouth of each bay (one point at the far-left side of the bay, another at the far-right side of the bay and the third point mid way of the two points), maximum effective fetch was calculated and the highest value taken to represent wave exposure in the bay. Bay dimensions were extracted by planimeter from topographic maps. The area of swamp in each bay was determined from vegetation maps and expressed as a fraction of total bay area. The resulting database on wave exposure and bay characteristics was used to investigate the variation of swamp area along a gradient of wave exposure through non-linear regression techniques.

Influence of exposure and bay features on the limit of vegetation advancement
The database was further used to assess, through hierarchical multiple linear regression, how well maximum effective fetch, fetch length variance, and the length, breadth, aspect ratio (i.e. ratio of bay width to breadth) and area of bays, predict the limit of lakeward vegetation advancement. To normalise score distribution, all variables were logarithm-transformed. Preliminary checks on data suitability identified weak to strong correlation between dependent and independent variables. However, strong correlation ($r > 0.9$) was also found between independent variables. To satisfy the assumption of multicollinearity, some variables (fetch length variance, bay length and bay breadth) were dropped and only three (wave exposure, bay area and aspect ratio) that were weakly correlated were retained for regression analysis.

For quantification of the limit of lakeward vegetation advancement, an estimate was required of the distance of the forward edge of vegetation from the land/water boundary. There was difficulty in determining this distance as the highly irregular shape of most of Lake Victoria's bays gives rise to multiple directions of vegetation advancement and leaves only the areas opposite the bay mouth under the direct influence of wind and waves from the open lake. This is illustrated in Figure 2.2. The problem was resolved by measuring the distance of the vegetation front relative to the bay mouth instead of the land/water boundary. For this purpose, it is assumed that wave action acts by limiting vegetation advancement: then, high wave action at the bay mouth would halt the advancement of vegetation. In this context, the vegetation boundary signifies a point of dynamic equilibrium between vegetation growth and destruction by wind and waves, other factors remaining favourable.

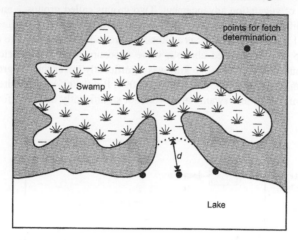

Figure 2.2. A hypothetical shoreline showing complex convolutions typical of the northern margins of Lake Victoria. The distance *d* of the vegetation front from the bay mouth was used as a surrogate for the limit of lakeward vegetation advancement.

Results

Vegetation distribution along the shore
The qualitative pattern of shoreline vegetation distribution on the northern shore of Lake Victoria suggests a strong influence of wave exposure on shoreline vegetation distribution. Shoreline plants are not randomly distributed: they occur in locations where the action of wind and waves is lowest. These typically are stretches of shore behind coastal islands and in backwaters. Wave exposure in vegetated parts of the shore ($\underline{M}$ = 23 $\pm$ 1.48 km) is significantly lower than in bare shores ($\underline{M}$ = 123 $\pm$ 5.54 km) (Mann-Whitney U test, z = -12.18, P < 0.001). However, there was considerable overlap in the wave exposure range of vegetated and non-vegetated shores (Figure 2.3). Wave exposure in vegetated shores had a positively skewed distribution, with lower exposure values occurring more frequently than higher exposure values.

Variation of swamp area along a gradient of wave exposure
The fraction of bay area covered by swamps was found to decrease exponentially with increasing wave exposure. There appeared to be some clustering in the bays which on scrutiny seemed to arise from differences in the range of within-site variability in fetch length. In the application of equation 2.1, fifteen values of fetch length are used to compute a single value of maximum effective fetch. Variability in fetch length can be small or large depending on the characteristics of the shoreline adjacent to the bay under consideration: it is small where the opposite shore is nearly straight; and large where the opposite shore is irregular, or where the presence of

islands causes abrupt discontinuity in long fetches. To examine the apparent signs of some kind of grouping in the bays, an hierarchical cluster analysis (between-groups linkage cluster method, Squared Euclidean distance measure, Z scores standardization) was performed using the variables wave exposure, bay area, aspect ratio and limit of vegetation advancement. Two groups called groups 1 and 2 were obtained that differed in the exponential coefficients for swamp area decline (Figure 2.4). In group 1, zero swamp cover is approached around a wave exposure of 33 km while in group 2, it is reached above a wave exposure of 140 km. Fetch length variance in group 1 ($\underline{M}$ = 2.3 $\pm$ 0.16 km) was significantly lower than in group 2 ($\underline{M}$ = 8.6 $\pm$ 0.23 km) (Mann-Whitney U test, z = -7.094; P < 0.0005). Other site features such as bay dimensions and swamp area were not significantly different for the two groups.

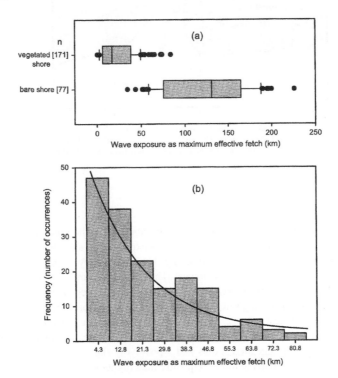

Figure 2.3. (a) Boxplots of wave exposure in bare and vegetated shores. The boundaries of each box indicate the 25th and 75th percentiles; the line in the box represents the median; the whiskers indicate the 10th and 90th percentiles and the black dots represent outliers. (b) Histogram showing wave exposure distribution in the sites occupied by shoreline vegetation.

Influence of exposure and bay features on the limit of vegetation advancement
Wave exposure, in combination with bay area and bay aspect ratio, explained 65.1% of the variance in the lakeward limit of vegetation progression (Table 2.1). Wave exposure and bay area were strongly and positively correlated with the limit of vegetation progression, and made statistically unique contributions (beta = 0.314 and

0.669 respectively) to the linear regression model. The model as a whole was significant [$F_{(3, 74)} = 46.09$; $P < 0.0005$]. The direction of relationships obtained shows that as wave exposure and bay area increase, the vegetation boundary recedes further and further away from baymouth into the bay. Bay aspect ratio was weakly correlated with the limit of vegetation advancement and made no unique contribution to the model.

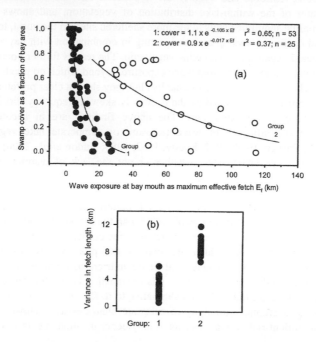

Figure 2.4. (a) Scatter diagram depicting the relationship between swamp cover and wave exposure in bays. The 78 bays were subdivided into groups 1 and 2 using hierarchical cluster analysis based on four variables. (b) Point plot of variance in fetch length in the two groups.

Table 2.1. Results of hierarchical multiple linear regression used to investigate the influence of wave exposure and coastal morphometric characteristics on the lakeward limit of vegetation progression.

No.	Model[a] Predictors	R Square	R Square change	F change	Sig. F change	Beta
1.	Constant, wave exposure	0.257	0.257	26.316	0.000	0.314
2.	Constant, wave exposure & bay area	0.644	0.387	81.570	0.000	0.669
3.	Constant, wave exposure, bay area & aspect ratio	0.651	0.007	1.527	0.221	0.092

[a] Dependent variable: limit of lakeward advancement

Discussion

Wave exposure as a regulator of shoreline vegetation distribution
This is the first quantitative demonstration of the applicability of the generalizations on the influence of wave exposure on shoreline vegetation distribution in a tropical lake. The results reinforce and amplify the hypothesis that wave exposure is a critical regulator of the within-lake distribution of vegetation, and shows wave-exposure to confine shoreline plants to the more sheltered sites on the coast, to cause an exponential decline in swamp area and, acting in combination with bay area, to limit the lakeward expansion of shoreline vegetation. The results further point to the existence of a yet unknown threshold wave exposure beyond which no plants occur. Below the threshold, plants are able to establish but they show a clear preference for low-exposure habitats as evidenced by the positively skewed frequency distribution of wave exposure in vegetated sites. The above findings are in accord with observations on the influence of wave exposure on temperate lake vegetation (Pearsall, 1920; Spence, 1964; 1967; Keddy, 1982, 1983; Hudon *et al.*, 2000). It was not possible from the data to state whether or not the lack of distinct growing seasons of tropical systems makes vegetation more resilient to disturbance by wind and waves than temperate vegetation.

Alternative explanations for the observed pattern of shoreline vegetation distribution may be offered but are less plausible than the action of wind and waves. The occupied sites might be more nutrient-rich than the un-occupied ones. Sediment deposition and resuspension, which are two of the processes responsible for producing variable substrate character and hence nutrient availability, are largely controlled by wind and wave action. Thus this explanation is an elucidation of the effect of wind and waves, rather than an alternative to it. Differences in geology and soil type may also produce variability in substrate characteristics. However, it is unnatural that nutrient-rich formations should only occur in embayments and behind islands.

Variable disturbance by man and animals could also lead to the observed pattern of shoreline vegetation distribution. However, the unfavourable physical environment of the shoreline wetlands (dense vegetation; boggy, unstable substrate; permanent inundation; mosquito and tsetse fly infestation) prohibits man and most large herbivores from entering the shoreline wetlands (Howard-Williams and Gaudet, 1985). A notable exception to the above generalisation is the hippopotamus, which grazes on swamp vegetation and causes considerable damage to trampled plants by its enormous body weight. Low dissolved oxygen and pH values, which are prevalent in the interior of shoreline wetlands, contribute to the creation of an inhospitable environment and hinder many fish and other aquatic organisms from utilising the interior of swamps. Thus, disturbance by man and animals may not be a significant contributor to the spatial distribution of shoreline vegetation.

The results of the study show a considerable overlap in the wave exposure range of bare and vegetated shores, which suggests that some habitable areas with respect to this disturbance are not colonized by vegetation. There are two possible reasons for this. The first is that the substrate in the uninhabited locations is not suitable for colonization by macrophytes. Spence (1967) from a study of Scottish lochs noted that shorelines comprised of boulders, coarse infertile gravels and highly reducing muds are rarely colonized by macrophytes. The second reason could be that the uninhabited locations have a high intensity or frequency of wave disturbance events. Further research is required to determine which of the two is the correct explanation.

Feedbacks between wave exposure and vegetation distribution
It has been shown (for the first time) that embayment area has a strong influence on the lakeward expansion of shoreline vegetation. The lakeward vegetation boundary was located much further back from baymouth in larger bays as compared to smaller bays. Since bay area is a primary determinant of the size of a bay's wind catchment, it is likely that the relationship obtained above is principally one between wave action and plant tolerance to this environmental control. It may be inferred from the relationship that in larger bays, a greater area of free-water surface is available for wave generation leading to correspondingly greater damage to plants.

If the above interpretation is correct, it implies that pioneer colonisers of a bay may facilitate secondary colonisation through a feedback system: the presence of pioneer plants reduces the available surface for wave generation, leading to reduced wind-wave activity, which in turn makes it possible for more plants to establish in the area. This constitutes a previously unrecognised positive feedback in the establishment of floating mats of vegetation, and might be a potentially stabilising mechanism for the development of an alternative vegetation state in lakes or embayments dominated by floating emergents (Scheffer *et al.*, 2003). Although the study emphasises that this relationship is likely complex, it is conceivable that feedback mechanisms of this nature play a role in the filling of bays by vegetation. Research to further examine feedbacks between wave exposure and plant establishment is recommended.

Disturbance intensity versus disturbance frequency
The difference that was found in the rate of swamp area decline in bays of differing fetch length variance suggests that it is not the maximal wave disturbance event in a year, but rather the overall stress levels caused by wave disturbance throughout the year, that have the greatest influence on shoreline vegetation development. In bays with small variance in fetch length, for a given wind speed, waves of about the same energy are generated from all compass directions used in effective fetch computation. On the other hand, in bays with large variance in fetch length, there are directions from which low-energy waves are generated as well as directions from which high-energy waves are generated. The resulting intermittent nature of strong wave action in the latter type of bays (assuming that winds are of variable direction during the year) may allow for periods of repair and recovery in between large disturbance events and produce the observed slower decline in vegetation cover with increasing wave exposure. Thus while surrogates of wave exposure like maximum effective fetch provide a simple and rapid means of estimating the magnitude of wave exposure, the results suggest that reliance on a measure of wave disturbance intensity without consideration for the disturbance frequency may lead to under- or over-estimation of the impact of the disturbance. This is especially so where there is a fairly even distribution of windy days in the calendar year. Coops *et al.*, (1991) reached a similar conclusion when studying the direct and indirect effects of wave action on shoreline vegetation.

Wind duration measures could not be incorporated in the estimation of wave exposure due to data limitations. In the study area there is one meteorological station located at Entebbe International Airport that measures hourly wind speed and direction. However, conditions at this single site could not be taken to represent conditions along the entire 675 km stretch of shore with its large variability in aspect, local topography and coastal morphometry.

Potential application in lake restoration
A potential exists for the use of exposure proxies, such as employed in this study, in restoration of degraded lakeshores. The methods could be particularly useful if appropriate factors to take into account the variability of wave disturbance frequency could be incorporated, as illustrated, for example, by Keddy (1982). A relationship between wave exposure and plant presence/absence or spatial extent, obtained from undisturbed parts of a lake, or from a nearby lake, could be used to identify locations along a degraded shore with a high potential for plant establishment. Since wave exposure is one environmental variable amongst many controlling the distribution of shoreline vegetation, the above techniques must of necessity be part of a suite of models covering the whole range of environmental regulating factors.

Acknowledgements

The field study was funded by the Lake Victoria Environmental Management Project (LVEMP) in Uganda. The assistance of Festus Rusoona in map work and fetch measurements is acknowledged.

References

Carpenter, S.R. (1983) Submersed macrophyte community strata and internal loading: relationship of lake ecosystem production and succession. *Lake and Reservoir Management* 2: 105-111.

CERC (Coastal Engineering Research Centre) (1984) *Shore Protection Manual.* Fort Belvoir, Virginia: U.S. Army Corps of Engineers.

Chambers, P.A. (1987) Nearshore occurrence of submersed aquatic macrophytes in relation to wave action. *Canadian Journal of Fisheries and Aquatic Science* 44: 1666-16669.

Coops, H., Boeters, R., and Smit, H. (1991) Direct and indirect effects of wave attack on helophytes. *Aquatic Botany* 41: 333-352.

Coops, H., Geilen, N., and Velde, v.d.G. (1994) Distribution and growth of the helophyte species *Phragmites australis* and *Scirpus lacustris* in water depth gradients in relation to wave exposure. *Aquatic Botany* 48: 273-284.

Crul, R.C.M. (1998) *Management and conservation of the African Great Lakes.* Paris: UNESCO.

Denny, P. (1973) Lakes of South-western Uganda II: vegetation studies on Lake Bunyonyi. *Freshwater Biology* 3: 123-135.

Denny, P. (1985a) Submerged and floating-leaved aquatic macrophytes (euhydrophytes). In *The ecology and management of African wetland vegetation.* Denny, P. (ed). Dordrecht: Dr. W. Junk Publishers, pp. 19-42.

Denny, P. (1985b) The structure and functioning of African euhydrophyte communities: The floating-leaved and submerged vegetation. In: P. Denny (Ed)., *The ecology and management of African wetland vegetation,* 125-152, Dr. W. Junk Publishers, Dordrecht.

Denny, P. (1985c) Wetland vegetation and associated life forms. In *The ecology and management of African wetland vegetation.* Denny, P. (ed). Dordrecht: Dr. W. Junk Publishers, pp. 1-18.

Denny, P. (1993) Eastern Africa. In *Wetlands of the World. I.* Whigham, D.F., Dykyjová, D., and Hejn , S. (eds). Dordrecht: Kluwer Academic Publishers, pp. 32-46.

Duarte, C.M. and Kalff, J. (1986) Littoral slope as a predictor of the maximum biomass of submerged macrophyte communities. *Limnology and Oceanography* 31(5): 1072-1080.

Foote, A.L. and Kadlec, J.A. (1988) Effects of wave energy on plant establishment in shallow lacustrine wetlands. *Journal of Freshwater Ecology* 4(4): 523-532.

Håkanson, L. and Jansson, M. (1983) *Principles of lake sedimentology.* Berlin: Springer-Verlag.

Howard-Williams, C. and Gaudet, J.J. (1985) The structure and functioning of African swamps. In *The ecology and management of African wetland vegetation.* Denny, P. (ed). Dordrecht: Dr. W. Junk Publishers, pp. 153-176.

Hudon, C., Lalonde, S., and Gagnon, P. (2000) Ranking the effects of site exposure, plant growth form, water depth and transparency on aquatic plant biomasses. *Canadian Journal of Fisheries and Aquatic Science* 57: 31-42.

Hutchinson, G.E. (1975) *A Treatise on Limnology.* New York: John Wiley and Sons.

Keddy, P.A. (1982) Quantifying within-lake gradients of wave energy: interrelationships of wave energy, substrate particle size and shoreline plants in Axe Lake, Ontario. *Aquatic Botany* 14: 41-58.

Keddy, P.A. (1983) Shoreline vegetation in Axe Lake, Ontario: effects of exposure on zonation patterns. *Ecology* 64: 331-344.

Keddy, P.A. (2000) *Wetlands - Principles and Conservation.* UK: Cambridge University Press.

Kennedy, C.W. and Bruno, J.F. (2000) Restriction of the upper distribution of New England cobble beach plants by wave-related disturbance. *Journal of Ecology* 88(5): 856-868.

Lind, E.M. and Morrison, E.S. (1974) *East African vegetation.* London: Longman Group.

LVEMP (Lake Victoria Environmental Management Project) (2002) *The Lake Victoria Integrated Water Quality and Limnology Study.* Consultant's Final Report, LVEMP Regional Secretariat, Dar-es-Salaam, Tanzania.

Maas, L.R.M., Benielli, D., Sommeria, J., and Lam, F.-P.A. (1997) Observation of an internal wave attractor in a confined stably stratified fluid. *Nature* 388: 557-561.

Pearsall, W.H. (1920) The aquatic vegetation of the English Lakes. *Journal of Ecology* 8: 163-201.

Riis, T. and Hawes, I. (2003) Effects of wave exposure on vegetation abundance, richness and depth distribution of shallow water plants in a New Zealand lake. *Freshwater Biology* 48: 75-87.

Scheffer, M., Szabo, S., Gragnani, A., van Nes, E.H., Rinaldi, S., Kautsky, N., Norberg, J., Roijackers, R.M.M., and Franken, R.J.M. (2003) Floating plant dominance as a stable state. *Ecology*, 100, 4040-4045.

Spence, D.H.N. (1964) The macroscopic vegetation of freshwater lochs, swamps and associated fens. In *The vegetation of Scotland.* Burnett, J.H. (ed). Edinburgh and London : Oliver and Boyd, pp. 305-345.

Spence, D.H.N. (1967) Factors controlling the distribution of freshwater macrophytes with particular reference to the lochs of Scotland. *Journal of Ecology* 55: 147-170.

Spence, D.H.N. (1982) The zonation of plants in freshwater lakes. *Advances in Ecological Research* 12: 37-125.

Floating mats: their occurrence and influence on shoreline distribution of emergent vegetation

Publication based on this chapter:

Nicholas Azza, Patrick Denny, Johan van de Koppel and Frank Kansiime (2006). Floating mats: their occurrence and influence on shoreline distribution of emergent vegetation. *Freshwater Biology* 51(7): 1286-1297. doi: 10.1111/j.1365-2427.2006.01565.x

Floating mats: their occurrence and influence on shoreline distribution of emergent vegetation

Abstract

A study was conducted on the northern shore of Lake Victoria (Uganda) to determine the factors controlling the occurrence of floating root mats and the influence of the floating mats on the distribution of emergent vegetation. Environmental conditions within 78 bays in the study area were characterized using bay size, wave exposure, water depth, littoral slope, sediment characteristics and water level fluctuations. Emergent plants that form floating root mats occur along the shores of these bays. The way in which commonly occurring shoreline vegetation was distributed across a wave exposure gradient was compared with their distribution across a water level fluctuation gradient. Results suggested that wind-wave action and water level fluctuations are important factors determining the occurrence of floating mats. Mat-forming plants occur in the most sheltered locations along the shore and in water bodies with low magnitudes of water level fluctuations. The mats appeared to facilitate the lakeward expansion of emergent plants. Plants forming floating root mats had a larger depth range than non-mat forming plants. The initiation mechanisms in Lake Victoria for the floating mats of emergent vegetation appear to be (1) invasion of the mats of free-floating plants by emergent vegetation; and (2) detachment of emergent plants from the lakebed following flooding. It is concluded that the formation of floating mats comes with a cost and benefit to emergent plants. The cost is increased vulnerability to damage by water level fluctuations or wind-wave action leading to reduced horizontal distribution. The benefit is avoiding the problem of deep flooding leading to increased vertical distribution. The net effect of the floating mat may be to cause dominance of mat-forming plants in low-energy environments and non-mat-forming plants in high-energy environments.

Keywords: **Floating mats, emergent vegetation, wave exposure, lake level fluctuation, Lake Victoria**

Introduction

The margins of many freshwater lakes and rivers are fringed by wetlands dominated by emergent herbaceous vegetation. Usually, the emergent vegetation is rooted in hydric mineral soils. In some cases, however, the emergent plants form and anchor in a buoyant mat that floats at a depth of a few centimetres from the water surface (Sculthorpe, 1967; Howard-Williams and Gaudet, 1985; Denny, 1993; Mitsch and Gosselink, 2000). The mat consists of live below-ground biomass, dead organic material and mineral sediments, and is held together by a meshwork of rhizomes and roots or the crisscrossing of stems (Azza et al., 2000). While free-floating at the lakeward end, the mat is secured by attachment in soil at the landward end. Its level rises and falls during the year in synchrony with seasonal fluctuations in the level of the adjacent river or lake (Swarzenski et al., 1991). The phenomenon of emergent mat formation has global geographical range (Mitsch and Gosselink, 2000), and commonly occurring emergent macrophytes such as *Phragmites* Trin., *Typha* L., *Scirpus* (Tourn.) L., *Carex* (Dill) L., *Cladium* P. Br., *Cyperus* L. and *Vossia* Wall. Et. Griff. all possess mat-forming ability (Denny, 1993).

The change from bottom-rooted to floating-mat growth forms, when it occurs, can have serious ecological and economic consequences making it imperative that factors regulating their development are known. With the aid of floating mats, dominant emergent macrophytes that are normally confined to the landward boundary of wetlands, start to extend towards open water, in the process shading and ultimately replacing euhydrophytes (Bernatowicz and Zachwieja; 1966). Below the mat, oxygen supply is poor and, commonly, hypoxic conditions prevail. The development of floating mats of emergent vegetation may hence be accompanied by a reduction in euhydrophyte abundance (as a consequence of shading), and in food and suitable habitat for fish and other aquatic fauna (as a result of decline in euhydrophytes and development of hypoxia). Furthermore, in unstable hydrological conditions, the forward ends of floating mats may break off and drift about. The free-floating vegetation islands formed in this manner can be a menace, littering recreation facilities and interfering with navigation and port activities (Sculthorpe, 1967; Terry and Minto, 1970; Thompson and Hamilton, 1983).

Despite the considerable potential to cause undesirable impacts and decrease the beneficial services of freshwater ecosystems to society, there is a poor appreciation of the factors that create and maintain floating root mats. It is not well understood why a given plant species should in one habitat grow rooted to the bottom and in another as a floating-mat. The way in which the mats arise and the benefits and constraints they confer upon emergent plants have not been fully investigated. Furthermore, there is no consensus in the literature on the ecological significance of floating emergent mats. Floating mat communities have been variously described as a stage in the succession from open water to dry land (Sculthorpe, 1967), an endpoint in succession characterised by remarkable stability in community structure (Sasser et al., 1995) and as an illustration of the adaptation by rhizomatous plants to problems of fluctuating water level (Keddy, 2000). The objective of this study is to improve understanding of the basic biology of floating mats particularly: (a) the factors controlling their development; (b) the mechanisms by which they arise; (c) their function; and (d) their influence on the shoreline distribution of emergent macrophytes.

Materials and Methods

Study area
The study was conducted in Lake Victoria in East Africa. The lake, the second largest in the world by surface area (68,870 km^2), has a maximum depth of 84 m, mean depth of 40 m and shoreline length of 3400 km (Beeton, 1984; Crul, 1998). Numerous bays occur along its shore, many fringed by tropical wetlands dominated by the emergent macrophytes *Cyperus papyrus* L. (a sedge), *Miscanthidium violaceum* (K. Schum.) Robyns (a grass), *Vossia cuspidata* (Roxb.) Griff. (also a grass) and *Phragmites mauritianus* Kunth (a reed). With the exception of *Phragmites*, the dominant emergents form floating mats, on which they and other shoreline wetland plants - mainly shrubs, herbs, ferns and mosses - anchor. *Phragmites* in the littorals of Lake Victoria occurs exclusively in the bottom-rooted form but forms floating mats elsewhere, as for example in the Danube Delta (Pallis, 1915; Rudescu et al., 1965). Accounts of the vegetation, structure and function of the shoreline wetlands of Lake Victoria are given by Lind and Morrison (1974), Howard-Williams and Gaudet (1985) and Denny (1993). The lake is fresh and eutrophic (LVEMP, 2002). A stretch of shore between Nabisukiro Channel and Berkeley Bay in the northern Ugandan part of the lake was selected for study (Figure 3.1). This stretch (510 km long) features convoluted and straight margins, sheltered and exposed shores, coastal archipelagos and fringe floating swamps of

various sizes. Seventy-eight (78) of the largest bays in this stretch with floating-mat swamps were selected for study.

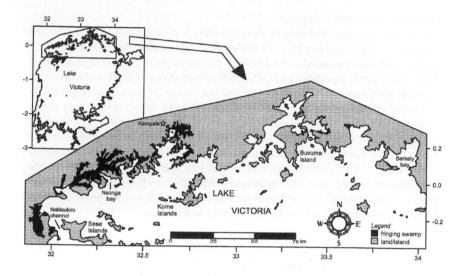

Figure 3.1. Map of the studied shore of Lake Victoria showing the distribution of coastal swamps. Map coordinates are in decimal degrees. Shoreline wetland cover was extracted from FAO Africover maps prepared from remote sensing images taken between 1999 and 2001.

Factors controlling floating mat development: approach

It was reasoned, that in a location where there is extensive development of emergent floating mats, it could be assumed that conditions favourable for mat establishment are prevalent. Therefore, it was considered that by examining the prevailing conditions in such a location, and by contrasting them with those in locations where emergent floating mats were absent, clues on the habitat requirements of emergent floating mats could be obtained. The study area, Lake Victoria, meets this need in having large areas of floating emergent swamps.

The abiotic environment within the selected bays was characterised using six variables. Four of the variables (bay area, wave exposure across bay mouth, water depth and littoral slope) were determined in all the bays. One variable, sediment character at the lakeward swamp edge, was determined for only a few cases while another variable, water level fluctuations, was evaluated at whole-lake scale. Statistical techniques were used to investigate the influence of bay area and wave exposure on the spatial extent of emergent floating swamps. The habitat requirement, with respect to wind and wave exposure, of emergent vegetation with the floating-mat growth form, was contrasted to that of bottom-rooted emergents by obtaining the distribution of common shoreline vegetation across a wave exposure gradient. In the literature, a review was found (Thompson, 1985) that compares the habitat requirement of floating and bottom-rooted emergent wetlands with regard to hydrological regime. Thompson's (1985) analysis was integrated with the results of the study in typifying the environment of floating-mat emergent macrophytes.

Features of the bays with floating mats

The area of each bay, and area of swamp within it were estimated from 1:50,000 topographic map sheets (series Y732 of Uganda) with a planimeter. Water depth and littoral slope were measured at the lakeward swamp edges in July 2004 using a boat fitted with a depth transducer and GPS. Duplicate readings of depth and slope were taken in each bay. At three points across the mouth of each bay (Figure 3.2) maximum effective fetch was calculated from 1:50,000 topographical map sheets using the method of Håkanson and Jansson (1983) and CERC (1984). Fetch, which is the open water distance over which wind-induced waves build, was used as a surrogate of wave-exposure. The northern shoreline of Lake Victoria is downwind of prevailing southerly winds. The influence of bay size and wave exposure on the proportion of swamp area was investigated through correlation with the Pearson product-moment coefficient of linear correlation. Bay size and exposure data was first logarithm-transformed to achieve normality and linearity.

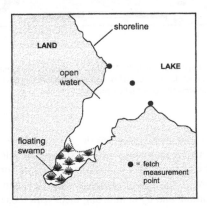

Figure 3.2. A hypothetical lakeshore showing a sheltered bay partially covered by a coastal swamp, and the points for fetch determination. The highest of the three fetch-determinations was taken to represent wave exposure in the bay.

At the time of littoral slope measurement, surficial (0-5 cm) sediment samples were collected from the lakeward edges of 10 randomly selected fringe swamps. For comparison, sediments from 10 nearshore areas off non-vegetated shores were sampled. Differences in median grain size, clay content, water content, organic matter content (determined as loss-on-ignition) and total phosphorus in the two groups of samples were assessed by Mann-Whitney U tests. Grain size distribution in samples was determined by sieving and hydrometer test (ASTM, 1986) and was used to calculate the clay content (proportion of particles finer than 0.063 mm) and median grain size (Håkanson and Jansson, 1983). Water content and loss-on-ignition were determined by drying and ignition respectively (APHA, 1995) while sediment phosphorus content was measured spectrophotometrically following ignition and extraction with boiling HCl (Andersen, 1976; APHA, 1995; Agemian, 1997).

Water level fluctuations
There are no records of hydrological measurements in the shoreline wetlands of Lake Victoria. Therefore, lake level data was obtained and analysed, this being taken as a proxy indicator of wetland hydroperiod. The Directorate of Water Development (DWD) in Uganda, which operates hydrological stations on the lake at Jinja and Entebbe, provided the data covering periods 1900 – 2004. Rainfall data for the island station of Bukasa, used to explore qualitatively the influence of the seasons, was also obtained from DWD. To obtain an idea of the relative magnitude of lake level fluctuations, the fluctuations of Lake Victoria were compared to that of three similarly sized great lakes Superior, Michigan-Huron and Erie using historical lake level data of the Department of Fisheries and Oceans, Canada (DFOC, 2005). Differences in within-year and between-consecutive-years level fluctuations in the lakes were assessed using one-way between-groups ANOVA with post-hoc (Tukey's HSD) tests.

Distribution of emergent vegetation along a wave-exposure gradient
A boat survey of a section of the shoreline (from Nsonga bay westwards to the areas south of Nabisukiro Channel; Figure 3.1) was conducted from 24-28 February 2003. During the survey, the locations of commonly occurring shoreline vegetation groups were marked with a GPS and coordinates later used to calculate maximum effective fetch (a proxy for wave exposure) for each location as above described. Specimens of dominant shoreline plants were collected and submitted to Makerere University Herbarium for identification.

Mat initiation by plant invasion from levées
The literature suggest four initiation mechanisms by which floating emergent mats are initiated. In the first, mat formation is promoted by the congregation of free-floating stoloniferous plants like *Pistia* L. and *Eichhornia* Kunth. on the water's edge. Detritus accumulated below the mat of the free-floating plants forms a favourable rooting medium for emergents, which then invade from the shoreline (Pallis, 1915; Sculthorpe, 1967; Somodi and Botta-Dukát, 2004). In the second mechanisms, emergent plants growing on levées start to expand towards open water (Russell, 1942; Sculthorpe, 1967; Gaudet, 1977). In the third mechanism, the upper layer of root material of emergent plants detaches from the lakebed and rises with attached mature plants to the water surface. The detachment can result from increasing buoyancy due to accumulating trapped gasses (Hogg and Wein, 1988a,b) or flooding (Clark and Reddy, 1998; Somodi and Botta-Dukát, 2004). In the fourth mechanism, a layer of un-vegetated substratum peels off from the lake bottom and rises to the water surface. Emergent plants subsequently establish on the mat from seeds and vegetative organs buried in the substratum (Clark and Reddy, 1998).

In this study, the possibility that the floating emergent mats of Lake Victoria arose through the second mechanism above is examined. This mechanism differs from the others in being sensitive to the texture and nutrient content of marginal soils. Under this hypothesis, pioneer colonizers commence their invasion of open water from levées, which must accordingly provide for their anchorage and mineral nutrient needs during the early stages of colonisation. In water bodies where this is the dominant mechanism of mat initiation, it is assumed that larger communities of plants with floating mats will establish adjacent to shores with soils of suitable texture and nutrient content. Therefore, it is expect that in such systems, soil type and the proportion of bay area covered by floating emergent swamps will be strongly correlated. Weak correlation, accordingly, implies that mat initiation occurs through alternative mechanisms. Correlation between soil type and floating swamp size was assessed with the Kruskal-Wallis test. Soil types along the margins of the selected bays were obtained from soil maps of Uganda prepared by Kawanda Agricultural Research Station. Classification of soils on the Kawanda maps is according to the USDA scheme (SSDA, 1993).

Results

Features of the bays with floating mats
Frequency distribution of bay size, swamp area and wave exposure were wide and positively skewed (Figure 3.3). Though the largest bay studied was 153 km^2 in size, half of the 78 bays were below 2 km^2. Similarly, the largest swamp encountered had an area of 85 km^2 but half the swamps were smaller than 0.6 km^2. The greatest degree of wave exposure at the mouth of swamp-fringed bays was an effective fetch of 115 km, but half the bays had effective fetch below 10 km. The depth of water at lakeward swamp edges was normally distributed within a range of 0.80 – 6.51 m (Figure 3.3). This depth range is much larger than has been reported for bottom-rooted emergents in temperate wetlands (Figure 3.4). The bed slope at swamp edges was very gentle with a mean of 0.033% and maximum of 0.168%. Sediment at vegetated and non-vegetated shores differed markedly: swamp-edge substrate were finer grained and had a significantly higher content of clay, water, organic matter and total phosphorus than substrate in non-vegetated shores (Table 3.1 and Figure 3.5). Differences in substrate character in the two types of sites investigated with the Mann-Whitney U test were significant for all parameters at the 0.01 level.

Exploration with the Pearson product-moment coefficient of linear correlation showed no correlation between bay size and the cover of floating emergent swamps ($r = -0.187$, $P = 0.100$, n = 78). One may reasonably assume small bays to be colonised and filled more easily than large bays. Contrary to this expectation, however, cases were found of small bays of size 1 km^2 and less having almost no vegetation, and large bays of size 20 km^2 and greater having more than half their surface area covered by floating emergent swamps. The corollary is that the size of bay per se does not determine the lakeward expansion of floating emergent vegetation. Wave exposure, on the other hand, had a noticeable effect on the size of floating emergent swamps. A moderate negative correlation ($r = -0.522$, $P < 0.005$) was found between wave exposure and swamp cover (Figure 3.6) indicating that emergent swamps occur where exposure to wind and waves is low. Since the littoral slope was evenly low everywhere, it could not contribute to the differential distribution of vegetation cover along the shore.

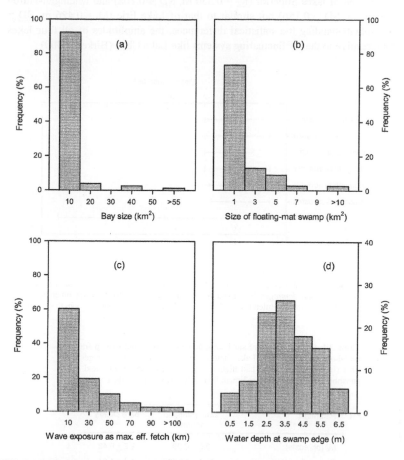

Figure 3.3. Histograms showing the distribution of bay size, swamp area, wave exposure and water depth in the study bays.

Water level fluctuations

With the exception of a single large water level rise in the 1960s, Lake Victoria's hydrological regime is reasonably stable, being characterised by modest within-year and between-consecutive-years water level fluctuations. Lake level rises and falls in a recurring sinusoidal pattern closely following rainfall (Figure3.7). Level fluctuations (within-year and between-consecutive-years) ranged from 0.01 to 1.66 m with a mean amplitude of 0.439 ± 0.017 m (Figure 3.8). The range of amplitudes of the other great lakes is comparable to that of Lake Victoria (Figure 3.8). However, there was a statistically significant difference at the $P<0.01$ level in the fluctuations [$\underline{F}$ (3, 688) = 23.01, P = 0.0005, eta squared = 0.09]. Post-hoc comparisons with Tukey's Honestly Significant Difference (HSD) test indicated the mean fluctuation of Lake Victoria ($\underline{M}$ = 0.439 m; $\underline{SD}$ = 0.249) to be significantly larger than that of lakes Superior ($\underline{M}$ = 0.330 m, $\underline{SD}$ = 0.102) and Michigan-Huron ($\underline{M}$ = 0.310 m, $\underline{SD}$ = 0.156) but similar to that of Lake Erie ($\underline{M}$ = 0.426 m, $\underline{SD}$ = 0.170). Notwithstanding the statistical differences, the amplitudes of all four lakes are small relative to that of fluctuating systems like Lake Chad (Birkett, 2000).

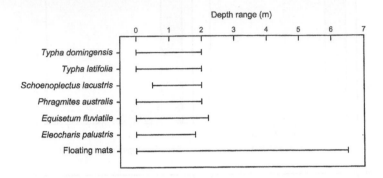

Figure 3.4. The depth distribution of emergent macrophytes growing along lakeshores. The indicated plants, with the exception of the floating mats, grow in the bottom-rooted form. The floating mats are a mixed community dominated by *Cyperus papyrus* and *Miscanthidium violaceum*. Data for *Typha domingensis* is from Welsh and Denny (1978). Data for other bottom-rooted (temperate) plants is from Hutchinson (1975) while data for floating mats is from this study.

Table 3.1. Characteristics of surficial sediments in bare and swamp-fringed nearshore environments of Lake Victoria. The sediment at swamp edges was finer (larger phi size) and had higher clay, water, organic matter and total phosphorous contents than sediment in un-vegetated parts of the coast.

Parameter	swamp-fringed shores	bare shores
M_d (phi scale)	9.5 $\pm$ 0.29	2.2 $\pm$ 0.43
Clay content (%)	61.3 $\pm$ 14.39	1.4 $\pm$ 1.84
Water content (%)	84.9 $\pm$ 1.65	21.3 $\pm$ 0.90
Organic matter as LOI (%)	52.0 $\pm$ 4.64	0.5 $\pm$ 0.09
TP (mg PO_4-P/mg ds)	1.63 $\pm$ 0.126	0.14 $\pm$ 0.014
n	10	10

The long-term hydrograph of Lake Victoria (Figure 3.9) shows a major anomaly between 1961 and 1964. Over a 30-months period between these dates, lake levels rose by 2.52 m. This anomaly has been attributed to unusually heavy rainfall in East Africa in 1961-1962 with corresponding increases in direct rainfall and tributary inflows to Lake Victoria (Flohn, 1987). Prior to this, levels fluctuated within a long-term range of 1.62 m. Post-1964 long-term level fluctuations are within a range of 1.21 m. Levels are slowly receding from the 1964 all-time-high at a rate of 0.032 m yr^{-1}.

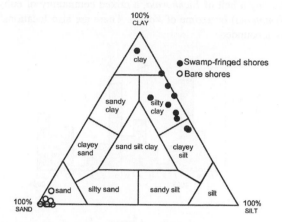

Figure 3.5. Textural classification of surficial sediments in bare and swamp-fringed nearshore environments of Lake Victoria. The coarse sandy material of bare shores contrasts markedly with the fine clayey substrate in vegetation fringed areas. The textural classification scheme is according to Shepard (1954).

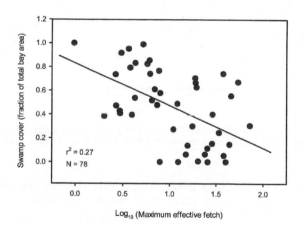

Figure 3.6. Scatter diagram showing the relationship between wave exposure and swamp cover in bays. There is a general decline in the proportion of bay area inhabited by emergent floating swamps as wave exposure within the bay increases.

Distribution of emergent vegetation along a wave-exposure gradient

The groups of vegetation encountered during the shoreline survey were: (a) floating papyrus and mixed papyrus/*Miscanthidium* swamps (with associated swamp plants); (b) the leguminous shrub *Sesbania sesban* L. which stands in up to 1.5 m of the water and is associated at the landward end with the low-growing perennial grass *Panicum subalbidum* Kunth.; and (c) the reed *Phragmites mauritianus* occurring exclusively in the bottom-rooted form and occupying sandy, shallow-flooded levées.

The floating papyrus/*Miscanthidium* swamps are commonly bounded on the lakeward end either by a belt of *Eichhornia*, a mixed community of euhydrophytes (dominated by *Nymphaea*) or a zone of *Vossia*. There are also locations where the floating swamp is unbounded.

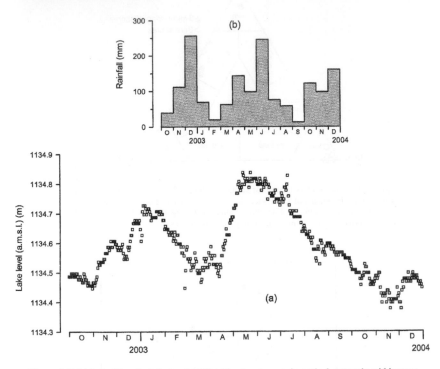

Figure 3.7. (a) Lake Victoria daily levels at Entebbe showing a twin-peaked pattern in within-year fluctuations that closely resembles (b) the rainfall pattern over the lake (measured on Bukasa Island).

The distribution of these plant groups along a wave exposure gradient (Figure 3.10) shows that bottom-rooted plants (both woody and herbaceous) occupy relatively exposed sites while floating-mat plants occupy the most sheltered sites on the coast. The difference in wave exposure in bottom-rooted ($\underline{M}$ = 47.83 km, $\underline{SD}$ = 17.24) and floating-mat ($\underline{M}$ = 14.64 km, $\underline{SD}$ = 12.54) sites of emergent macrophytes compared with an independent-samples t-test was significant ($\underline{t}$ (134) = -11.237; P =0.0005; eta squared = 0.485). The bottom-rooted phragmites (occupying sites with a fetch range of 14-84 km) had a wider distribution range with respect to wave exposure than floating papyrus/*Miscanthidium* swamps (fetch range 0.5-45 km). The obligate acropleustophyte *Eichhornia* comes to rest only in the most sheltered sites. These results suggest that macrophytes with the floating-mat growth form are more vulnerable to wind-wave disturbance than bottom-rooted macrophytes. A positive impact of the floating mat on emergent plants is also potentially possible. An x-y plot combining the distribution of emergent plants (of different genera) across wave exposure and depth gradients (Figure 3.11) indicates that in well-sheltered habitats, emergent vegetation achieves greater spatial cover when occurring in the floating mat form.

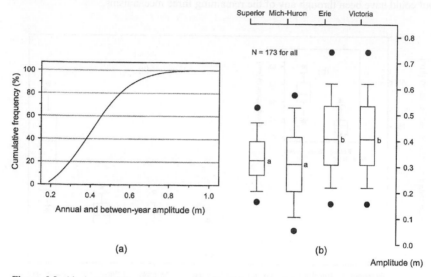

(a)

(b)

Amplitude (m)

Figure 3.8. (a) A cumulative frequency plot of within-year and between-consecutive-years level fluctuations for Lake Victoria for the period 1900-2004. (b) Boxplots comparing within-year/between-consecutive-years amplitudes of lakes Superior, Michigan-Huron, Erie and Victoria (for the period 1918-2004). Lakes with the same symbol have similar amplitudes and differ significantly from those with a different symbol. The boundaries of each boxplot indicate the 25th and 75th percentiles; the line in the box represents the median; the whiskers indicate the 10th and 90th percentiles and the black dots indicate the 5th and 95th percentiles.

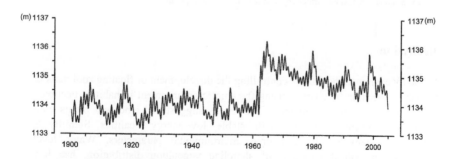

Figure 3.9. Lake Victoria annual minimum and maximum levels at Entebbe

Mat initiation by plant invasion from levées

Soils on the landward margin of the bays studied were of four types: sand (n = 7), sandy loam (n = 41), sandy clay loam (n = 19) and sandy clay (n = 11). There was no clear relationship between the marginal soil types and fractional area of floating emergent swamps in the bays (Figure 3.12; Kruskal-Wallis test, Chi-Square = 4.181, df = 3, asymptotic significance = 0.243). The inference was that mat initiation did not follow mechanism two (i.e. lakeward expansion of emergent plants from levées) but could have been through any of the remaining three mechanisms.

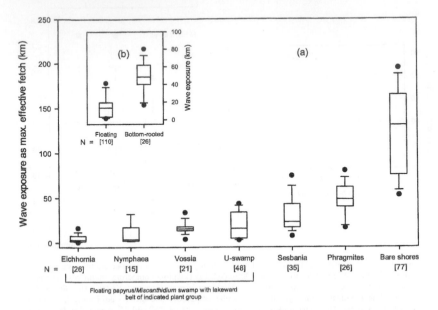

Figure 3.10: Boxplot showing: (a) the ranges on an exposure gradient of generic vegetation groups occurring on the northern shoreline of Lake Victoria; and (b) the distribution range on an exposure gradient of floating-mat (papyrus and Miscanthidium) and bottom-rooted (phragmites) emergents. The group labelled U-swamp represents floating papyrus and Miscanthidium swamps unbounded by other vegetation. See text for further explanations on vegetation groups.

Discussion

This study examined the factors controlling the development of floating root mats in emergent vegetation by characterising the abiotic environment in habitats occupied by macrophytes with floating mats. The results suggest that emergent plants with floating mats occur mostly in small, stable, low-energy localities. Stability is apparent in many of the measured environmental parameters. Water level fluctuations, a critical regulator of shoreline vegetation distribution, has low amplitude; bays occupied by emergent macrophytes with floating mats are mostly small and well-sheltered (i.e. low wind-wave exposure); and embayments fringed by floating swamps have fine-grained sediments rich in clay content, water content and organic matter, which is typical of low-energy environments.

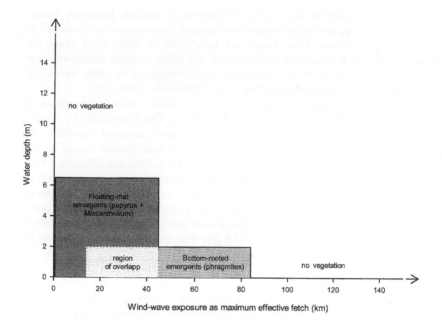

Figure 3.11. An x-y plot showing the distribution ranges of bottom-rooted and floating-mat emergent plants across wave exposure and water depth gradients.

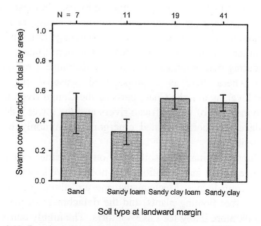

Figure 3.12. Bar chart showing the swamp-cover in bays of different marginal soil type. Soil types from left to right are arranged in order of increasing textural fineness and fertility. Error bars indicate the standard error of the mean.

The results of this study suggest that the adoption of the floating-mat growth form imposes two habitat requirements on emergent macrophytes. The first is shelter from wind and wave disturbance. While all shoreline plants are affected by wind and wave action (Keddy, 1982; Foote and Kadlec, 1988; Coops *et al.*, 1991), plants with floating mats seem more vulnerable to this disturbance than bottom-rooted plants.

Stability with respect to water level fluctuations is another important habitat requirement that appears to be imposed on emergent macrophytes by the switch to the floating-mat growth form. Large and rapid rises in water level cause the peripheral ends of floating mats to break away and either establish elsewhere when low-water conditions return, or get crushed and disintegrated by wind, waves and water currents (Thompson and Hamilton, 1983; Thompson, 1985).

The relative stability of lake level may partially explain the extensive development of floating swamps around Lake Victoria. This is conferred by a fairly even distribution of rainfall over the year and small seasonal variability in evaporation. A number of studies clearly show rainfall to be a major control on the lake's water balance (see for example Piper *et al.*, 1986; Sene and Plinston, 1994; Yin and Nicholson, 1998). Geostrophic and meteorological factors have a comparatively smaller influence on the lake's water level fluctuations than seasonal change. Coriolis force is considered negligible as the lake straddles the equator (Talling and Lemoalle, 1998). Meteorologically-induced seiches cause random oscillations of water level at periods of 3.5 h (the first resonant harmonic of the lake) and amplitudes in the order of 100 mm (GOU, 1997).

While stability is undoubtedly an important requirement, it is unlikely that this factor alone is responsible for the extensive development of floating swamps, since comparable stability also prevails in the Laurentian great lakes that have no floating emergents. It is hereby speculated that significant development of floating mats might occur where favourable environmental factors coincide with the optimal altitudinal and latitudinal distribution range of vigorous mat-forming emergents.

These results agree with the conclusions of Thompson (1985) about the habitats of floating-mat and bottom-rooted emergent macrophytes. Thompson (1985) noted that in aquatic systems where the hydrological regime is stable, as in the Upemba swamps of the Democratic Republic of Congo, the Okavango delta in Botswana, the Kafue floodplains in Zambia and the Sudd area of southern Sudan, shallow levées are occupied by floating-mat emergents commonly dominated by papyrus. In the contrasting system where there is a pronounced seasonal regime and large amplitudes in water level, as in Lake Chad, parts of the Congo river, the Chambeshi river in Zambia, the majority of west African rivers and the lower Nile, floating-mat emergents are replaced by bottom-rooted emergents, commonly but not always dominated by phragmites.

The lack of influence of marginal soil types on the spatial extent of emergent floating swamps suggests that mat initiation does not involve lakeward expansion or emergent vegetation from levées. Mechanisms one and three (i.e. the invasion by emergents, of mats of free-floating plants, and the detachment of mature emergents from the lakebed) look more likely for Lake Victoria. The highly convoluted margin of the lake has many backwaters that are ideal resting places for free-floating plants and, hence, potential nuclei for mat formation. The lake level rise of the 1960s could also have led to detachment of emergent plants from the lakebed as it caused extensive flooding of low-lying marginal lands, including shoreline wetlands (Azza *et al.*, 2000). Moreover, reconstructions of lake level changes in the nineteenth century (Nicholson and Yin, 2001) and over the past 1,100 years (Verschuren *et al.*, 2000) have revealed that the 1960s hydrological event was not unique. The level of Lake Victoria (and of other East African lakes) has been alternating between extreme low and high stands in response to large episodic shifts in regional climate. Thus, it would seem that some of the floating swamps of Lake Victoria have arisen from recurrent flooding of bottom-rooted shoreline wetlands.

The flood-mediated mechanism would seem more applicable to *Miscanthidium* than papyrus. *Miscanthidium* has a compact, closely-knit mat that contrasts markedly with the more open, loosely bound mat of papyrus. Additionally, the lower reaches of the *Miscanthidium* mat are almost impermeable (Azza *et al.*, 2000), which would make them more efficient at trapping gaseous products of microbial mineralisation activities, and enhancing mat buoyancy.

What deductions can be made about the role of floating root mats? The better-known functions of the mat are in related to its role in colonisation and succession (Sculthorpe, 1967) and in nutrient cycling and water purification (Howard Williams and Gaudet, 1985; Kipkemboi *et al.* 2002). Keddy (2000) has additionally suggested that mats may provide a means by which emergent vegetation avoid problems of fluctuating water level. However, such a function is contradicted by the field distribution pattern of floating-mat emergents presented above. A possible previously unrecognised function of the floating mat is in enabling emergent plants to overcome the limitation of water depth to lakeward expansion.

In the classical distribution of vegetation along the depth continuum from dry land to open water (Hutchinson, 1975; Spence, 1982; Mitsch and Gosselink, 2000), emergent plants are confined to a narrow strip at the water's edge due mainly to their inability to withstand prolonged deep flooding, but also to competition from neighbours. With the aid of the floating root mat, emergent vegetation is able to colonise areas of deepwater without investing in physiological adaptations for deep flooding. In theory, plants with floating root mats should be able to colonise any depth of water giving the plants the potential to cover entire lakes. In practice, emergent plants with floating mats are not distributed beyond the nearshore zone (except for drifting pieces detached from parent mats). This, I believe, is largely due to the destructive effect of wind and waves in open waters than to any inability of the mats to float over deep water.

It is clear that, while the adoption of the floating root mat benefits emergent plants (i.e. through increased vertical distribution), it also has a cost (i.e. an increased vulnerability to water level and wind-wave disturbance leading to reduced horizontal distribution). For a given river or lake system, the mat's net influence is likely to depend on the balance between stable low-energy environments and unstable high-energy environments in the totality of locations available for plant colonisation. Where the gain in spatial cover through greater vertical distribution is larger than the loss from reduced horizontal distribution, mat formation could lead to increased abundance of mat-forming emergents. However, the lakeward expansion of emergent vegetation inevitably means displacement and reduced abundance of euhydrophytes. Thus the floating root mat may be an important phenomenon influencing the structure of shoreline wetlands in aquatic systems where floating-mat emergents occur.

The above-mentioned distributions across wave exposure and hydrological regime gradients are ecological response curves of bottom-rooted and floating-mat emergents. In the absence of neighbours and competition, both plant groups may distribute differently across these environmental gradients. Research to compare distributions with and without competition is therefore recommended. Wave exposure and hydrological regime are just two of many regulators of the shoreline distribution of aquatic vegetation. Further research is required to establish how the presence of the floating mat influences the impact of the numerous abiotic and biotic controls operating in the natural habitat of emergent vegetation.

References

Agemian, H. (1997) Chapter 7: Determination of nutrients in aquatic sediments. In *Manual of physico-chemical analysis of aquatic sediments*. Mudroch, A., Azcue, J.M., and Mudroch, P. (eds). Boca Raton, Florida: Lewis Publishers, CRC Press Inc., pp. 175-227.

Andersen, J.M. (1976) An ignition method for determination of total phosphorous in lake sediments. *Water Research* 10: 329-331.

APHA (American Public Health Association) (1992) In *Standard methods for the examination of water and wastewater*. American Public Health Association, pp. 2-57 to 2-65.

ASTM (American Society for Testing and Materials) (1986) Standard method for particle-size analysis of soils: Method D 422. In *Annual book of ASTM standards*. Vol. 04 Section 4: Construction: 08: Soil and rock; building stones Philadelphia: American Society for Testing and Materials, pp. 116-126.

Azza, N.G.T., Kansiime, F., Nalubega, M., and Denny, P. (2000) Differential permeability of papyrus and *Miscanthidium* root mats in Nakivubo swamp, Uganda. *Aquatic Botany* 67(3): 169-178.

Beeton, A.M. (1984) The world's Great Lakes. *Journal of Great Lakes Research* 10(2): 106-113.

Bernatowicz, S. and Zachwieja, J. (1966) Types of littoral found in the lakes of the Masurian and Sulwaki lakelands. *Ekologiya Polska* 14(Ser. A): 519-545.

Birkett, C.M. (2000) Synergistic remote sensing of lake Chad - variability of basin inundation. *Remote Sensing of Environment* 72(2): 218-236.

CERC (Coastal Engineering Research Centre) (1984) *Shore Protection Manual*. Fort Belvoir, Virginia: U.S. Army Corps of Engineers.

Clark, M.W. and Reddy, K.R. (1998) *Analysis of Floating and Emergent Vegetation Formation in Orange Lake* . St. Johns River Water Management District, Florida.

Coops, H., Boeters, R., and Smit, H. (1991) Direct and indirect effects of wave attack on helophytes. *Aquatic Botany* 41: 333-352.

Crul, R.C.M. (1998) *Management and conservation of the African Great Lakes*. Paris: UNESCO.

Denny, P. (1985) Wetland vegetation and associated life forms. In *The ecology and management of African wetland vegetation*. Denny, P. (ed). Dordrecht: Dr. W. Junk Publishers, pp. 1-18.

Denny, P. (1993) Eastern Africa. In *Wetlands of the World. I.* Whigham, D.F., Dykyjová, D., and Hejn , S. (eds). Dordrecht: Kluwer Academic Publishers, pp. 32-46.

DFOC (Department of Fisheries and Oceans Canada), Fluctuations in lake levels: long-term (multi-year), retrieved on: 14 February 2005, from: http://chswww.bur.dfo.ca/danp/fluctuations_e.html#TC11

Flohn, H. (1987) East African rains of 1961/62 and the abrupt change of the White Nile discharge. *Paleoecology of Africa* 18: 3-18.

Foote, A.L. and Kadlec, J.A. (1988) Effects of wave energy on plant establishment in shallow lacustrine wetlands. *Journal of Freshwater Ecology* 4(4): 523-532.

Gaudet, J.J. (1977) Uptake, accumulation and loss of nutrients by papyrus in tropical swamps. *Ecology* 58: 415-422.

GOU (Government of Uganda)(1997) *Lake water hydrodynamic studies in the Murchison bay area of Lake Victoria*. Final Report by HR Wallingford in Association with Gibb (Eastern Africa). Government of Uganda, Kampala.

Hakanson, L. and Jansson, M. (1983) *Principles of lake sedimentology*. Berlin: Springer-Verlag.

Hogg, E.H. and Wein, R.W. (1988) The contribution of Typha components to floating mat buoyancy. *Ecology* 64(4): 1025-1031.

Hogg, E.H. and Wein, R.W. (1988) Seasonal change in gas content and buoyancy of floating Typha mats. *Journal of Ecology* 76(4): 1055-1068.

Howard-Williams, C. and Gaudet, J.J. (1985) The structure and functioning of African swamps. In *The ecology and management of African wetland vegetation*. Denny, P. (ed). Dordrecht: Dr. W. Junk Publishers, pp. 153-176.

Hutchinson, G.E. (1975) *A Treatise on Limnology*. New York: John Wiley and Sons.

Keddy, P.A. (1982) Quantifying within-lake gradients of wave energy: interrelationships of wave energy, substrate particle size and shoreline plants in Axe Lake, Ontario. *Aquatic Botany* 14: 41-58.

Keddy, P.A. (2000) *Wetlands - Principles and Conservation*. UK: Cambridge University Press.

Kipkemboi, J., Kansiime, F., and Denny, P. (2002) The response of *Cyperus papyrus* (L.) and *Miscanthidium violaceum* (K. Schum.) Robyns to eutrophication in the natural wetlands of Lake Victoria, Uganda. *African Journal of Aquatic Science* 27(1): 11-20.

Lind, E.M. and Morrison, E.S. (1974) *East African vegetation*. London: Longman Group.

LVEMP (Lake Victoria Environmental Management Project) (2002) *The Lake Victoria Integrated Water Quality and Limnology Study*. Consultant's Final Report, LVEMP Regional Secretariat, Dar-es-Salaam, Tanzania.

Mitsch, W.J. and Gosselink, J.G. (2000) *Wetlands*. New York: John Wiley and Sons.

Nicholson, S.E. and Yin, X. (2001) Rainfall conditions in equatorial East Africa during the nineteenth century as inferred from the record of Lake Victoria. *Climate Change* 48: 387-398.

Pallis, M. (1915) The structural history of Plav: the floating fen of the delta of the Danube. *Journal of the Linnaean Society, Botany* 43: 233-290.

Piper, B.S., Plinston, D.T., and Sutcliffe, J.V. (1986) The water balance of Lake Victoria. *Hydrological Sciences Journal* 31(1): 25-37.

Rudescu, L., Niculescu, C., and Chivu, J.P. (1965) *Monografia Stufului din Delta Dunarii* . Bucharest: Editura Academiei Republicii Socialiste Romania.

Russell, R.J. (1942) Floatant. *Geographical Review* 32: 74-78.

Sasser, C.E., Visser, J.M., Evers, D.E., and Gosselink, J.G. (1995) The role of environmental variables on interannual variation of species composition and biomass in a subtropical minerotrophic floating marsh. *Canadian Journal of Botany* 73: 413-424 .

Sculthorpe, C.D. (1967) *The Biology of Aquatic Vascular Plants*. London: Edward Arnold.

Sene, K.J. and Plinston, D.T. (1994) A review and update of the hydrology of Lake Victoria in East Africa. *Hydrological Sciences Journal* 39(1): 47-63.

Shepard, F.P. (1954) Nomenclature based on sand-silt-clay ratios. *Journal of Sedimentary Petrology* 23(3): 151-158.

Somodi, I. and Botta-Dukát, Z. (2004) Determinants of floating Island vegetation and succession in a recently flooded shallow lake, Kis-Balaton (Hungary). *Aquatic Botany* 79(4): 357-366.

Spence, D.H.N. (1982) The zonation of plants in freshwater lakes. *Advances in Ecological Research* 12: 37-125.

SSDA (Soil Survey Division Staff) (1993) *Soil Survey Manual*. Washington D.C.: Soil Conservation Service. U.S. Department of Agriculture. U.S. Government Print Office.

Swarzenski, C.M., Swenson, E.M., Sasser, C.E., and Gosselink, J.G. (1991) Marsh mat floatation in the Louisiana Delta Plain. *Journal of Ecology* 79: 999-1011.

Talling, J.F. and Lemoalle J. (1998) *Ecological dynamics of tropical inland waters*, Cambridge University Press, Cambridge.

Terry, P.J. and Minto, J.D. (1970) *Floating Islands in Lake Victoria*. Report, Tropical Pesticides Research Institute (TPRI), Arusha, Tanzania .

Thompson, K. (1985) Emergent plants of permanent and seasonally-flooded wetlands. In *The ecology and management of African wetland vegetation*. Denny, P. (ed). Dordrecht: Dr. W. Junk Publishers, pp. 43-108.

Thompson, K. and Hamilton, A.C. (1983) Peatlands and swamps of the African continent. In *Mires: Swamp, Bog, Fern and Moor*. Gore, A.J.P. (ed). Amsterdam: Elservier Scientific Publishing company, pp. 331-373.

Verschuren, D., Laird, K., and Cumming, B.F. (2000) Rainfall and drought in equatorial East Africa during the past 1,100 years. *Nature* 403: 410-414.

Welsh, R. and Denny, P. (1978) The vegetation of Nyumba ya Mungu reservoir, Tanzania. *Biological Journal of the Linnaean Society* 10: 67-92.

Yin, X. and Nicholson, S.E. (1998) The water balance of Lake Victoria. *Hydrological Sciences Journal* 43(5): 789-811.

Chapter 4

Could changing lake conditions explain littoral wetland and fish species decline in Lake Victoria?

Submission to a scientific journal based on this chapter:

Nicholas Azza, Patrick Denny, Steven Loiselle, Luca Bracchini, Andres Cozar, Arduino Dattilo and Johan van de Koppel. Could changing lake conditions explain littoral wetland and fish species decline in Lake Victoria (East Africa)? *Canadian Journal of Fisheries and Aquatic Sciences,* submitted on October 12, 2006.

Could changing lake conditions explain littoral wetland and fish species decline in Lake Victoria?

Abstract

Recently, the number of fish species and their populations in Lake Victoria dropped markedly. The main theory put forward to explain these changes – Nile perch predation – cannot account for all aspects of the fish population upheaval making it necessary to find other causal factors. This chapter examines the possibility that littoral wetland degradation, as a consequence of eutrophication, led to the degradation of littoral fish habitats and, hence, became partly responsible for the observed decline in fish species. To investigate this hypothesis, modern physico-chemical and optical properties from five nearshore areas in Lake Victoria are compared with past conditions. At a sixth study area, diel changes in lake condition were followed over a 24-hour cycle. Data confirmed the ongoing eutrophication of the lake. Areas bordering fringe wetlands were contrasted from other parts of the lake in having lower dissolved oxygen, pH and Secchi depth values, and higher chlorophyll-a, dissolved organic matter and light attenuation values. On a daily basis in wetland-fringed areas, there is a recurrent rise and fall in un-ionised ammonia concentrations in rhythm with the diel cycles of water temperature, pH and dissolved oxygen. Changing lake conditions, it is argued, may have gradually altered the survival chances of fish exploiting marginal habitats and differentially influenced their decline in speciation and numbers through (1) causing a loss of littoral vegetation, especially euhydrophytes, through increased light attenuation leading to loss of food, breeding grounds and refugia; and (2) causing the development of unfavourable conditions characterized by hypoxia and high ammonia concentrations. This suggests a strong possibility that the loss of fish biodiversity in Lake Victoria was caused in part by elimination of marginal refugia that had protected fish from excessive Nile perch predation.

Keywords: Lake Victoria, wetland degradation, cichlids, species decline, eutrophication.

Introduction

Lake Victoria, the world's second largest freshwater body by area, had until recently over 500 species of fish, the greater majority of which were haplochromine cichlids (Greenwood, 1974, 1981; Witte *et al.*, 1992). Starting in the early 1980s a dramatic decline occurred in the fish populations with over 200 species of endemic cichlids becoming extinct (Ogutu-Ohwayo, 1990; Witte *et al.*, 1992). It is important that the causal factors and mechanisms for this catastrophe are clearly identified as such knowledge could form the basis for efforts aimed at preventing further extinctions and promoting faunal recovery.

Initially, it was considered that intense predation by Nile perch was the overriding cause of the faunal collapse (Ogutu-Ohwayo, 1990: Witte *et al.*, 1992). The decline in endemic fish populations coincided with a major upsurge in the population of the Nile perch, a non-native piscivore introduced in Lake Victoria in 1954. This view was partially borne out by observations of resurgence in cichlid populations after commercial over-fishing moderated Nile perch predation pressure (Witte *et al.*, 2000). However, the hypothesis cannot account for the disappearance of cichlids that are rarely eaten by Nile perch, such as stenotopic rock-dwelling species, and for habitat shifts in surviving species (Seehausen, 1996; Seehausen *et al.*, 1997a,b). Hence it is now widely accepted that predation alone cannot suffice as an explanation for the recent changes in fish populations, and that other contributing factors need to be considered (Kaufman and Ochumba, 1993; Bundy and Pitcher, 1995; Seehausen *et al.*, 1997a; Witte *et al.*, 2000).

One possible causal factor for the fish biodiversity loss that has been suggested but not intensively investigated is the loss of habitat following degradation of littoral wetlands as a result of eutrophication (Balirwa, 1995). Lake Victoria underwent major transformation in its ecosystem from around 1960 to 1990 that is largely attributed to cultural eutrophication (Hecky, 1993; Lung'ayia et al., 2001). It is possible that changes in the physical and chemical condition of the lake brought about by eutrophication caused a loss of habitat for many species of fish and quickened their disappearance.

Several reasons can be given as to why such a view is worthy of further consideration. First, the body of literature clearly indicates that littoral zones, which commonly have a rich floral and faunal diversity and hence abundant food supply, function as important fish feeding, spawning and nursery grounds (Craig, 1987; Stephenson, 1990, Jude and Pappas, 1992; Brazne and Beals, 1997). Turner (1977) has shown that the yields of coastal fisheries worldwide are highly positively correlated with littoral wetland area, suggesting a tight link between nursery ground function of shoreline wetlands and the size of adult fish populations. Within Lake Victoria itself, marginal wetlands have been noted for their abundance of juveniles and diversity of fish species (Kaufman and Ochumba, 1993; Mnaya and Wolanski, 2002). Second, shoreline wetlands have been shown to provide fauna with valuable structural and physiological refuges from predation (Burks et al., 2001; Laegdgaard and Craig, 2001). Fish in vegetated shoreline habitats avoid capture by hiding amongst the dense foliage of aquatic macrophytes and retreating to areas within shoreline wetlands that are relatively dark (impairs visual predation) and hypoxic (keeps out predators not able to tolerate low-oxygen conditions). Recent studies have established that over 50% of the cichlid species thought to have gone extinct in Lake Victoria are still extant in small satellite water bodies around the lake where structural and physiological refuges are well developed (Chapman et al., 1996, 2002, 2003; Mwanja et al., 2001; Aloo, 2003). Third, theory backed by extensive empirical evidence suggests that in aquatic systems dominated by submerged macrophytes, eutrophication causes the development of turbidity and a shift to a phytoplankton-dominated state (Moss, 1988; Scheffer, 1998; Carpenter et al., 1999). Such change is possible in Lake Victoria in the interface zone between shoreline vegetation and open water. This zone is normally inhabited by euhydrophytes (a collective term for submersed, floating-leaved and bottom-rooted aquatic macrophytes; Denny, 1985c), which are the most important plants with respect to the support functions to fish. The shift to a phytoplankton-dominated state that is triggered by eutrophication is usually accompanied by changes in community structure (Scheffer et al., 1993). Euhydrophytes disappear or decline in spatial cover and species diversity, and fish and other fauna feeding on the plants or on invertebrates associated with the plants perish. The loss of vegetation also means loss of refuge and change in predator-prey relations for many organisms (Scheffer and van Nes, 2004).

Given the above functions of littoral wetlands, it seems likely that qualitative or quantitative changes in their condition could have important implications for the survival of cichlids. However, owing to a paucity of monitoring data, it is not possible to test this hypothesis by, for example, comparing chronological reconstructions of changes in abundance and species composition of littoral vegetation and wetland-dwelling fish with predictions of theory.

In this study modern conditions in nearshore waters of Lake Victoria are compared with historical (early 1960s and 1990s) conditions. Several physico-chemical and optical characteristics are combined in the investigation, the combination serving as a proxy indicator of the state of the lake. The objective is to assess if recent changes in Lake Victoria's condition were of a type with a potential to cause littoral wetland degradation and, hence, alteration of the chances for survival of fish utilizing marginal habitats for feeding, breeding and refuge. An affirmative outcome would imply the possibility that littoral wetland degradation contributed to fish species decline. Related to the above goal, potential mechanisms by which the interaction of eutrophication with marginal wetlands could lead to decline in fish species are explore.

Materials and Methods

Study areas and scope of study
Modern conditions that were contrasted with the past state of Lake Victoria were obtained from field measurements carried out in June 2003 at six nearshore study areas in the northern part of the lake (Figure 4.1). The areas range from well-sheltered wetland-fringed embayments to exposed wave-disturbed nearshore areas (Table 4.1). The emergent macrophytes *Cyperus papyrus* L. and *Miscanthidium violaceum* (K. Schum.) Robyns, both of which form floating root mats on which plants anchor, dominate floral composition in the wetland-fringed bays (Lind and Morrison, 1974). At the interface between emergent swamps and open water is usually a zone of euhydrophytes commonly dominated by water lilies, pondweed and eelgrass (Lind and Visser, 1962; Denny, 1985b).

At five of the six areas studied (Kagegi, Katonga, Nsonga, Lido and Tende), ambient physico-chemical and optical conditions were determined, and at the sixth (Berkeley), diurnal changes in physico-chemical condition were followed over one 24-hour cycle. At the five study areas, daytime levels of pH, dissolved oxygen (DO), chlorophyll-a, dissolved organic matter (DOM), Secchi depth transparency and light attenuation coefficients were measured. Within each study area, measurements were taken at several points roughly distributed on a 1 x 1 km grid. At the sixth study area (Berkeley) vertical profiles of water temperature, electrical conductivity (EC), pH, oxidation-reduction potential, dissolved oxygen, ammonium, total ammonia and nitrate were measured every 30 minutes. Detailed descriptions of methods are given below.

Water quality profiles
A Hydrolab Datasonde 4a (HACH Environmental Inc., Loveland) was used to measure vertical profiles of water temperature, electrical conductivity, pH, oxidation-reduction potential, dissolved oxygen and ammonium, total ammonia and nitrate. At the five study areas (where only pH and DO were measured), readings, which were taken at one-second intervals during descent, were processed to give depth-averaged values of each parameter. At the sixth study area (where all parameters were measured), readings were processed to give depth-time distributions of each parameter. The Hydrolab measures concentrations of

ammonium ion (NH_4^+) using an ion-selective electrode and calculates corresponding concentrations of un-ionized ammonia gas (NH_3) from the equilibrium relationship between NH_4^+ and NH_3, and *in vivo* concentrations of NH_4^+. Effects of ambient water temperature, pH and ionic strength on the position of the equilibrium are incorporated in the calculations. By subtracting the ammonium ion concentration from total ammonia results, the concentration of un-ionized ammonia corresponding to each ammonium measurement was obtained.

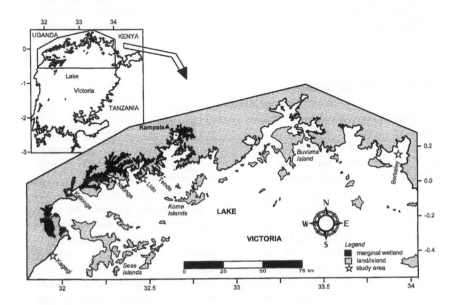

Figure 4.1. Map of northern Lake Victoria showing study areas. Map coordinates are in decimal degrees. Shoreline wetland cover was extracted from FAO Africover maps prepared from remote sensing images taken between 1999 and 2001.

Chlorophyll-a

Water was sampled with a Van Dorn horizontal water sampler at a depth of 0.25 m, and the natural fluorescence of samples was measured *in situ* with a SCUFA II Fluorometer (Turner Designs Inc., Sunnyvale). Natural fluorescence values were converted into chlorophyll-a concentrations using a calibration curve prepared from simultaneous fluorescence and chlorophyll-a analyses on a set of lake samples. Chlorophyll-a concentrations for the calibration curve were obtained spectrophotometrically after sample filtration through Whatman GF/F glass microfibre filter paper and extraction with 100% ethanol (Pápistra *et al.*, 2002).

Table 4.1. Characteristic features of study areas.

Study area	Area (km²)	Mean Depth (m)	Sample Points (No.)	Features
Katonga	18.6	2.2	22	Well-sheltered, large fringe wetland, river inflow
Nsonga	18.1	4.2	25	Well sheltered, large fringe wetland at landward end
Tende	12.3	5.2	15	Sheltered, small fringe wetland, some rocky margins
Lido	13.5	7.5	16	Exposed, no fringe wetland, wide sandy beach
Kagegi	22.7	4.6	10	Highly exposed, no fringe wetland, wide sandy beach
Berkeley	161	7.2	1	Partially exposed, small fringe swamp, river inflow

Dissolved organic matter (DOM)

Part of the sample collected at 0.25m as described above was filtered through a 0.22 µm Millipore membrane filter and stored in the dark at 4 °C till analysis (on same day). Preliminary analyses had shown that DOM concentrations did not change significantly in the 0-2 m layer. Upon return from the field, sample absorbance was measured at 272 nm with a UV/Visible Ultrospec 2000 spectrophotometer (Amersham Pharmacia Biotech, Buckinghamshire) against a distilled water blank. Samples were read in quartz cuvettes with a 0.01 m path length. The wavelength of 272 nm allows the quantification of all forms of dissolved organic matter in water (Mazzuoli *et al.*, 2004). Humic and fulvic acids as well as semi-stable decomposition products of organic compounds absorb strongly at this wavelength. Sample absorbance was converted into DOM concentrations from a calibration curve prepared with a commercial humic acid formulation (Sigma-Aldrich Chemie GmbH, Taufkirchen) comprising 49% humic acids (molecular weight >30,000 g mol^{-1}), 12% fulvic acids (molecular weight 5000 – 30,000 g mol^{-1}) and 39% low-molecular-weight DOM (< 5000 g mol^{-1}) (Mazzuoli *et al.*, 2004).

Secchi disc transparency and light attenuation

Secchi disc transparency was conventionally measured by lowering a black-and-white disc into the water until it disappeared from view. Vertical profiles of cosine-corrected downwelling irradiance of UV (305, 313, 320 and 340 nm) and Photosynthetically Active Radiation (PAR: 400-700 nm) were measured with a submersible PUV-541 Profiling Ultraviolet Radiometer (Biospherical Instruments Inc., San Diego). Prior to field use, the instrument was calibrated using a SUV-100 high-resolution UV/Visible Scanning Spectroradiometer (Biospherical Instruments Inc., San Diego). All light measurements were made between 10:00 and 15:00 hours on days with clear blue skies and little wind. At each measurement point, the profiler was deployed via a pulley set 2 m clear of the boat to avoid interference from the boat's shadow, and was slowly lowered to the lake bottom and hoisted back to the surface. Data was captured at 0.3 s intervals during descent and ascent. Irradiance values used to correct for dark current signal were determined by taking measurements with the sensor head covered with a light-tight neoprene cap (Kirk,

1994). In the water column, downwelling irradiance decreases exponentially with depth (due to scattering and absorption) according to the Beer-Lambert law:

$$I_\lambda = I_{o,\lambda} \exp(-K_{d,\lambda}[z-z_o]) \qquad (4.1)$$

where I_λ (Watts m^{-2} nm^{-1}) is the irradiance at depth Z (m), $I_{o\lambda}$ (Watts m^{-2} nm^{-1}) is the irradiance at depth Z_o (m) (= 0.02 m in this study) and $K_{d,\lambda}$ (m^{-1}) is the vertical extinction coefficient for downwelling light of wavelength λ (nm). A linear form of equation (4.1) was used to determine $K_{d,\lambda}$ by linear regression techniques:

$$\mathrm{Log}\left(\frac{I_{o,\lambda}}{I_\lambda}\right) = \frac{K_{d,\lambda}}{2.303}(z-z_o) \qquad (4.2)$$

Fits for which the coefficient of determination (R^2) was less than 0.95 were discarded.

Results

Ambient physico-chemical and optical conditions
Modern physico-chemical and optical conditions are indicative of a eutrophic state in the lake. Although conditions in the five study areas are non-uniform and the wetland-fringed areas show large within-study area variability in character (Figures 4.2 and 4.3), generally, Secchi disc transparency is low while phytoplankton pigment concentrations are moderate to high in the lake's margin habitats. There were locations, in all five study areas, with saturated to supersaturated oxygen concentrations and strongly alkaline pH. These are classic indicators of eutrophication (Harper, 1992). In the figures, the 313 nm band is used to depict the pattern of change in downwelling UV light. In a related study (Bracchini *et al.*, in press), the vertical profiles and between-study area variability in the other bands of UV light were found to be similar to the 313 nm band. It is considered, therefore, that this band provides a good indication of the vertical changes in the whole UV range.

Data was obtained from literature that provide an idea of nearshore physico-chemical and optical conditions in the early 1960s and 1990s. Comparison of the historical and modern data (Table 4.2) shows modern conditions to be markedly different from conditions in the 1960s. The comparison suggests the rate of change to be fairly rapid as conditions in the early 1990s - just over a decade ago - are noticeably different from present day conditions.

There are perceptible between-site trends in physico-chemical and optical conditions. Moving from bare-shore, through intermediate to wetland-fringed bays, there is a gradual decrease in DO, pH and Secchi depth and increase in the concentrations or values of chlorophyll-a, DOM, $K_{d,PAR}$ and $K_{d,313}$. Thus wetland-fringed study areas have lower DO, pH and Secchi depths, and higher chlorophyll-a, DOM, $K_{d,PAR}$ and $K_{d,313}$ than the bare-shore study areas. The differences between the five bays for the combined parameters were found to be statistically significant (one-way between-groups MANOVA $F[28, 240] = 7.78$, $P = 0.0005$; Pillai's Trace $= 1.90$; partial eta squared $= 0.48$). Further tests on between-subjects effects (by fitting of a general linear model) revealed that the bays differ significantly ($P <$ 0.01) with respect to all parameters except for DO.

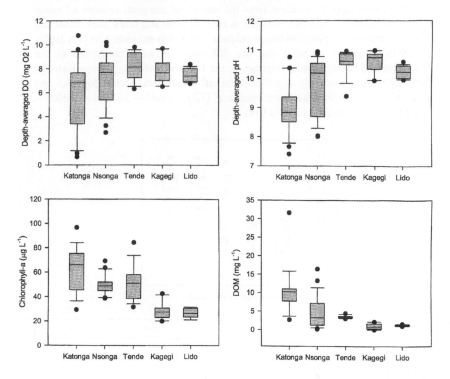

Figure 4.2. Boxplots showing modern conditions with respect to DO, pH chlorophyll-a and DOM in the five nearshore areas of Lake Victoria. The lower and upper borders of each box indicate the 25th and 75th percentiles; the line in the box represents the median; the whiskers indicate the 10th and 90th percentiles and the black dots indicate outliers. Well-sheltered wetland-fringed areas had large variability in parameter values while exposed areas had little variability. The general trend from wetland fringed to bare-shore is an increase in DO and pH, and decrease in chlorophyll-a and DOM.

Bare-shore habitats, as mentioned above, have little variability in physico-chemical and optical conditions and are akin to open lake waters (see for example LVEMP (2002) for characteristics of offshore waters). Wetland-fringed bays, on the other hand, have large variability in physico-chemical and optical conditions (Figures 4.2 and 4.3). Further investigation of the variability in conditions through gridding and contouring found no obvious patterns in bare-shore study areas but clear patterning in wetland-fringed bays. In the latter, there is a systematic gradation in water character moving from the bay/open-lake interface, through middle parts of the bay to the wetland edge (Figure 4.4). Points at the bay/open-lake boundary have high levels of dissolved oxygen, strongly alkaline pH, low concentrations of DOM, HA and chlorophyll-a, high transparency and low light attenuation. The water character at the bay/open-lake interface is comparable to that in bare-shore bays and in the open lake. Points located close to marginal swamps are of opposite character: they typically have low levels of dissolved oxygen, neutral pH, high concentrations of DOM, high concentrations of chlorophyll-a, low transparency, and high light attenuation. Points in the middle parts of the bay have transitional character. These results seem to suggest an outflow from the interior of marginal swamps of water that is nearly devoid of dissolved oxygen, is rich in humic and fulvic substances and is of acidic reaction. A mixed between-subjects and within-subjects ANOVA (split-plot ANOVA design) revealed that the differences between conditions at the swamp-edge and bay/open-lake interface in the two wetland-fringed bays Katonga and Nsonga are statistically significant for all parameters except for DO and chlorophyll-a (ranges for significant results: F (1,7) = 11.87 – 58.8; P = 0.0005 – 0.01; partial eta squared = 0.629 – 0.894).

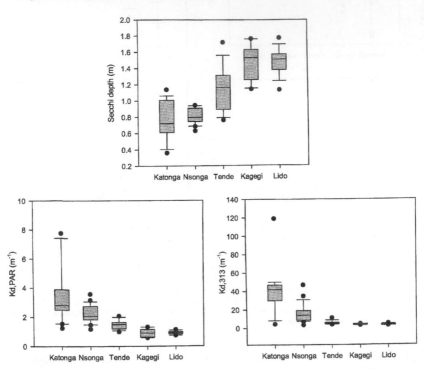

Figures 4.3. Boxplots showing modern conditions with respect to Secchi depth, $K_{d,PAR}$ and $K_{d,313}$ in the five nearshore areas of Lake Victoria. Secchi disc transparency, as expected from theory, varies in an opposite manner to light attenuation coefficients. Moving from wetland-fringed to bare-shore areas, Secchi depth increases while $K_{d,PAR}$ and $K_{d,313}$ decrease.

Diel changes in physico-chemical conditions

The diurnal data (Figure 4.5) shows a rhythmic rise and fall in water column temperature as the lake heats up from insolation by day and cools down by advection by night. The diel temperature cycle is closely mimicked by cycles in water column DO and pH presumably reflecting waxing and waning in phytoplankton activity (Talling, 1966, Talling and Lemoalle, 1998).

Table 4.2. Modern and historical physico-chemical and optical conditions in nearshore areas of northern Lake Victoria. The values in the table are ranges (in brackets) and means (in front of brackets).

Period (and site of measurement)	Chlorophyll-a (μg L^{-1})	Secchi depth (m)	$K_{d,PAR}$ (m^{-1})	$K_{d,313}$ (m^{-1})
1960-61 (Pilkington Bay)[1]	12.5 (10-15)	-	0.5 (0.16-0.62)[4]	0.59 (0.42-0.83)[4]
1989-91 (Pilkington Bay)[2]	46.7 (22.2-67.1)	1.2 (0.8-1.7)	1.02 (0.81-1.26)	-
2003 (Five northern bays)[3]	47.2 (20.1-96.6)	1.0 (0.3-1.8)	1.95 (0.58-7.76)	16.77 (2.98-119.0)

[1] Source: Talling (1965; 1966)
[2] Source: Mugidde (1993)
[3] This study
[4] Data is for offshore station (Bugaia) in northern Lake Victoria

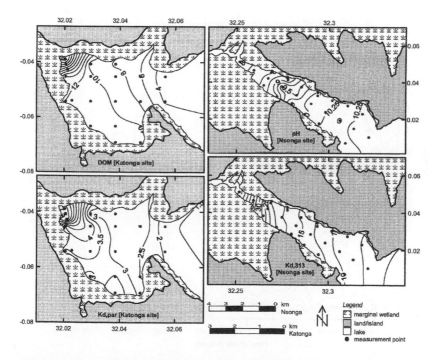

Figures 4.4. Spatial variation of DOM (mg L^{-1}), pH and attenuation coefficients for visible ($K_{d,PAR}$; m^{-1}) and UV ($K_{d,313}$; m^{-1}) light in wetland-fringed nearshore areas of Lake Victoria. Other physico-chemical parameters not shown had similar spatial distribution patterns. Map coordinates are in decimal degrees.

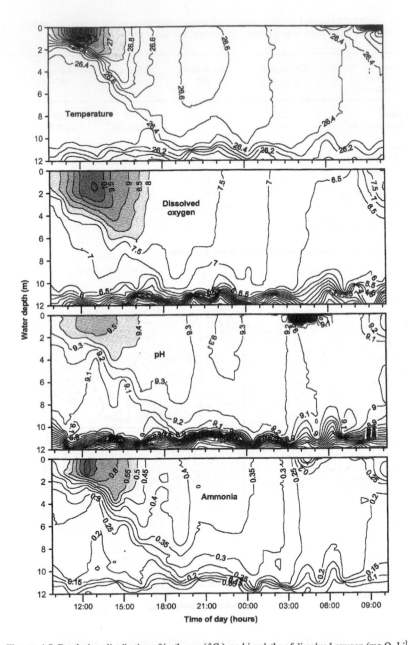

Figures 4.5. Depth-time distribution of isotherms (°C), and isopleths of dissolved oxygen (mg O_2 L^{-1}), pH and un-ionised ammonia (mg NH_3-N L^{-1}) at the Berkeley site, Lake Victoria during June 22-23 2003. The daily peak in all parameters (grey-filled region) occurs around midday at the time of strongest insolation.

Quite strangely, total ammonia and un-ionised ammonia were found to fluctuate in rhythm with changes in water temperature (Figure 4.5). The rest of the parameters (electrical conductivity, ORP, NH_4^+ and NO_3^-) do not have clear diel patterns.

Discussion

Changes in the condition of Lake Victoria

In this study, modern conditions in nearshore waters of Lake Victoria were compared with historical (1960s to early 1990s) conditions. This was done to assess if recent changes in condition were of a type with a potential to modify littoral wetlands in such a way as to alter the chances of survival of fish utilizing marginal habitats for feeding, breeding and refuge.

Modern physico-chemical and optical conditions in Lake Victoria, relative to historical conditions, are strikingly different. In areas adjacent to marginal wetlands DOM and phytoplankton pigment concentrations are high, PAR and UV light penetration is low, Secchi disc transparency is low, dissolved oxygen concentrations are low and un-ionised ammonia is produced in surface waters around the time of strongest insolation. These changes in lake condition, it is believed, have a potential to degrade euhydrophyte-dominated littoral habitats and impair the functioning of marginal wetlands as fish habitats.

A possible interfering influence on the physico-chemical and optical condition of nearshore waters is the contribution of groundwater through the base flow component of influent streams. If the quality of groundwater in the catchment is significantly different from the quality of surface or rainwater, it may introduce differences in nearshore conditions between areas adjacent to large inflowing streams and areas away from the streams. However, the influence of groundwater in the lake is negligible. The Lake Victoria region is made up predominantly of old crystalline rocks from the Pre-Cambrian era (3000 – 6000 million years ago) and has poor ferralitic soils that have lost nearly all of their mineral content through prolonged weathering (Taylor and Howard, 2000). Measurements of physico-chemical conditions along a 200 km stretch of shoreline between Katonga and Murchison Bays (Cózar *et al.*, 2004) showed that the quality of nearshore water adjacent to inflowing streams is indistinguishable from the quality of water away from inflowing streams except where the stream is heavily polluted by activities in the catchment (mainly the discharge of untreated municipal effluent).

Littoral wetland degradation

Light is the single most important factor regulating the shoreline distribution of submerged plants (Spence, 1982). Increase in light attenuation could easily lead to disappearance of plant species unable to elevate their leaves to the shrinking photic zone or hold their leaves up at the lake surface. Thus, it is possible that the marked decline in underwater light penetration that was observed above could have been accompanied by qualitative and quantitative changes in the euhydrophyte community of Lake Victoria.

Some change has certainly occurred in the littoral vegetation of Lake Victoria, for the extensive macrophyte beds and high floral diversity reported by previous workers were not found during the fieldwork. Early studies of the littoral swamps of Lake Victoria found their lakeward edges to be dominated by *Vossia cuspidata* (Roxb.) Griff., an emergent grass, and by several rooted, floating-leaved and submerged species, namely *Nymphaea lotus* L., *N. heudelotii* Planch., *N. caerulea* Sav., *Trapa natans* L., *Brasenia peltata* F. Pursh and *Nymphoides (Limnanthemum) nilotica* (Kotschy & Peyr.) Léonard. An understorey dominated by single or mixed communities of *Ottelia ulvifolia* (Planch.) Walp., *Ceratophyllum dermersum* L., *Utricularia thonningii* Schum., *U. foliosa* L., *Vallisneria spiralis* L. (Eur), *Potamogeton schweinfurthii* A. Benn. and *P. thunbergii* Cham. & Schlecht. was also common (Eggelling, 1935; Lind and Visser, 1962; Denny, 1985b). The modern patches appear to be somewhat smaller and less diverse (Azza, personal observation).

In Lake Victoria, the role of light in regulating the occurrence of euhydrophytes is probably more critical than in other lakes. This is because euhydrophyte vegetation of the lake are in competition for shallow marginal areas with floating emergent plants (Denny, 1985a; Azza *et al.*, in prep). Commonly, floating emergents take over shallow landward areas and, in so doing, shade out or edge euhydrophytes into deeper water. With their photosynthetic organs positioned deep below the lake surface, euhydrophytes are likely to be eradicated by any change that leads to reduced light penetration. This is true for submersed and rooted growth forms as well as floating-leaved growth forms. While mature floating-leaved plants are not limited by light attenuation under water, their seedlings, which occupy a depth range of about 0.5 to 2.0 m require underwater light for photosynthesis and continue to do so until their leaves come to the surface. Thus, increased light attenuation ultimately leads to complete loss of the euhydrophyte zone.

Potential mechanisms linking eutrophication, littoral wetlands and fish species decline
There are three main ways, in the author's view, by which fish could have been affected by the recent transformation in the condition of Lake Victoria. The first, which has been suggested in the literature (Seehausen, 1997a), is through a lake-wide reduction in underwater visibility, which interferes with mate choice and sexual selection and prevents reproductive isolation - a necessary condition for species diversity. The second, mostly affecting wetland-dwelling fish, is through the loss of habitat and hence food, breeding ground and refuge, following the decline or disappearance of euhydrophytes as a consequence of reduced underwater light penetration. For fish utilising littoral wetlands for feeding, breeding and refuge, the presence of euhydrophytes in the wetland is of utmost importance. The third, also mostly affecting wetland-dwelling fish, is through the development of hostile and stressful environmental conditions characterized by hypoxia, strongly alkaline pH and high ammonia concentrations.

Through the above means, the change in lake condition could have led to a decline or extinction of fish species unable to find the right sexual partners in poor light conditions, unable to shift to new habitats after loss of original habitats or unable to withstand hostile and stressful environmental conditions resulting from eutrophication. Conversely, the changed conditions could have enhanced the survival of those species best able to adapt to these changes, particularly those able to withstand the stressful conditions at the edge or interior of floating emergent vegetation. The Nile perch, which has a low tolerance for hypoxia and other physiologically stressful conditions (Schofield and Chapman, 2000), is known to avoid the edges and interior of marginal wetlands. Thus, the interaction of eutrophication with marginal wetlands could partially explain Lake Victoria's differential decline in fish species, and habitat shifts in surviving fish species.

The processes responsible for light attenuation appear to be an increase in phytoplankton biomass and washout of chromophoric dissolved organic matter from fringe wetlands. In Lake Victoria, the increase in algal biomass with concomitant decrease in water transparency in response to eutrophication has been well documented (Mugidde, 1993) and seems, from the above comparison of phytoplankton pigment levels, to be on the increase. Washout of DOM from fringe swamps also seems to be on the rise. This could be due to increasing land clearance and deforestation in the catchment (Hamilton, 1984; Cohen et al., 1996; Crul, 1998) leading to greater runoff and flushing of the fringe swamps by inflowing streams. There have been no previous measurements of the content of DOM in Lake Victoria but Talling (1966) commented that the concentrations of "yellow substances" (i.e. humic and fulvic compounds which are a component of DOM) appeared to be rather low in the early 1960s.

Dissolved organic matter in the marginal habitats affects light levels by directly absorbing UV/Visible light and supporting dense growths of algae, which cause further shading. When DOM is present in high concentrations, it removes nearly all of the incoming solar radiation within the first few decimetres of the water column (Jones, 1992; Carpenter et al., 1998; Scully and Lean, 1994; Morris et al., 1995). The daytime rise in NH_3 at the Berkeley study area (Figure 4.5) is probably due to the reversible conversion of NH_4^+ to NH_3 following its photochemical production from DOM. The nitrogen component of DOM is to some extent bioavailable and is degraded in shoreline waters by both bacterial (Carlson and Granéli, 1993; Carlson et al., 1993; Seitzinger and Sanders, 1997) and photochemical processes (Bushaw et al., 1996; Bushaw and Moran, 1999; Zagarese et al., 2001). Photochemical degradation has been found to lead to the formation of small, biologically labile organic compounds, and to the production of primary amines (Amador et al., 1989; Jørgensen et al., 1998), free ammonium (Bushaw et al., 1996; Gardner et al., 1998; Bushaw and Moran, 1999) and urea (Jørgensen et al., 1998).

Since Lake Victoria is mainly considered to be nitrogen limited (Talling, 1966, Hecky, 1993; Lehman and Brandstrator, 1993; Mugidde, 1993; Holtzman and Lehman, 1998) the formation of inorganic N from DOM represents an important supplementary source of nitrogen for phytoplankton productivity and may be responsible for the higher levels of algal biomass in wetland-fringed nearshore areas as compared to bare-shore areas. Hence, high levels of DOM in shoreline habitats may cause a decline in euhydrophyte vegetation and fish populations through their direct and indirect effects on light attenuation.

A further way in which DOM could affect fish is through the production of free ammonium as a by-product of DOM photobleaching. Ammonia, which is toxic to fish, exists in aqueous solution in two forms: the ammonium ion NH_4^+ and un-ionised dissolved ammonia gas (NH_3), with relative concentrations of the two forms being dependent on pH, water temperature and ionic strength (Emerson et al., 1975).

The un-ionised NH_3 form, which is predominant when pH is above 9.7, is more toxic than the ionized biologically-labile NH_4^+ form (Downing and Merkens, 1955). When ammonia levels are high in water, fish tend to accumulate ammonia which ultimately causing convulsions and death of the fish (Randall and Tsui, 2002). Thus the presence of DOM and high solar irradiance in shoreline wetland habitats may combine to produce unfavourable conditions that force fish to migrate to other habitats or abandon the relatively security of macrophyte beds for the more habitable but predator-prone region of offshore waters. This process may be exacerbated by nutrient-rich runoff from agricultural land and inflow of untreated sewage, which encourage greater biomass production and hence detritus and humic-substance formation in shoreline swamps.

Suggestions for management and further research
It was not possible to more than speculate upon possible changes in the community structure of shoreline wetlands because quantitative data detailing systematic changes in spatial cover and floristic composition over the years is not available for any marginal wetland in Lake Victoria. This limitation makes difficult the substantiation of the above hypothesis with empirical evidence. Accordingly, the commencement of periodic monitoring of the littoral wetlands as a necessary means of improving understanding of the dramatic changes occurring in the fish populations of Lake Victoria is recommended. Even more important than field monitoring is the protection of littoral vegetation from further degradation given their important ecological functions to fish and the lake ecosystem as a whole.

Acknowledgements

This research was funded by the Lake Victoria Environmental Management Project (LVEMP) and the European Commission DG RTD INCO DEV Programme (ICA4CT2001-10036).

References

Aloo, P.A. (2003) Biological diversity of the Yala swamp lakes, with special emphasis on fish species composition, in relation to changes in the Lake Victoria Basin (Kenya): threats and conservation measures. *Biodiversity and Conservation* 12: 905-920.

Amador, J.A., Alexander, M., and Zika, R.G. (1989) Sequential photochemical and microbial degradation of organic molecules bound to humic acid. *Applied Environmental Microbiology* 55: 2843-2849.

Azza, N., Denny, P., and Koppel, v.d.J. What factors control the development of floating mats of emergent vegetation? In prep.

Balirwa, J.S. (1995) The Lake Victoria environment: Its fisheries and wetlands - a review. *Wetlands Ecology and Management* 3(4): 209-224.

Bracchini, L., Cózar, A., Dattilo A.M., Loiselle, S.A., Tognazzi, A., Azza, N., and Rossi, C. (In press) The role of wetlands in the chromophoric dissolved organic matter release and its relation to aquatic ecosystems optical properties. A case of study: Katonga and Bunjako Bays (Victoria Lake, Uganda). *Chemosphere.*

Brazner, J.C. and Beals, E.W. (1997) Patterns in fish assemblages from coastal wetland and beach habitats in Green Bay, Lake Michigan: a multivariate analysis of abiotic and biotic forcing factors. *Canadian Journal of Fisheries and Aquatic Sciences* 54: 1743-1761.

Bundy, A. and Pitcher, T.J. (1995) An analysis of species changes in Lake Victoria: did the Nile Perch act alone? In *The Impact of Species Changes in African Lakes*. Pitcher, T.J. and Hart, P.J.M. (eds). London: Chapman and Hall, pp. 111-135.

Burks, R.L., Jeppesen, E., and Lodge, D.M. (2001) Littoral zone structures as *Daphnia* refugia against fish predators. *Limnology and Oceanography* 42(6): 230-237.

Bushaw, K.L., Zepp, R.G., Tarr, M.A., Schulz-Jander, D., Bourbonniere, R.A., Hodson, R.E., Miller, W.L., Bronk, D.A., and Moran, M.A. (1996) Photochemical release of biologically labile nitrogen from aquatic dissolved organic matter. *Nature* 381: 404-407.

Bushaw, K.L. and Moran, M.A. (1999) Photochemical formation of biologically available nitrogen from dissolved humic substances in coastal marine systems. *Aquatic Microbial Ecology* 18: 285-292.

Carlsson, P. and Granéli, E. (1993) Availability of humic bound nitrogen for coastal phytoplankton. *Estuarine, Coastal and Shelf Science* 36: 433-447.

Carlsson, P., Segatto, A.Z., and Granéli, E. (1993) Nitrogen bound to humic matter of terrestrial origin - a nitrogen pool for coastal phytoplankton? *Marine Ecology Progress Series* 97: 105-116.

Carpenter, S.R., Cole, J.J., Kitchell, J.F., and Pace, M.L. (1998) Impact of dissolved organic carbon, phosphorous, and grazing on phytoplankton biomass and production in experimental lakes. *Limnology and Oceanography* 43: 73-80.

Carpenter, S.R., Ludwig, D., and Brock, W.A. (1999) Management of eutrophication for lakes subject to potentially irreversible change. *Ecological Applications* 9(3): 751-771.

Chapman, L.J., Chapman, C.A., and Chandler, M. (1996) Wetland ecotones and refugia for endangered fishes. *Biological Conservation* 78: 263-270.

Chapman, L.J., Chapman, C.A., Nordlie, F.G., and Rosenberger, A.E. (2002) Physiological refugia: swamps, hypoxia tolerance and maintenance of fish diversity in the Lake Victoria region. *Comparative Biochemistry and Physiology Part A* 133: 421-437.

Chapman, L.J., Chapman, C.A., Schofield, P.J., Olowo, J.P., Kaufman, L., Seehausen, O., and Ogutu-Ohwayo, R. (2003) Fish fauna resurgence in Lake Nabugabo, East Africa. *Conservation Biology* 17(2): 500-511.

Cohen, A.S., Kaufman, L., and Ogutu-Ohwayo, R. (1996) Anthropogenic threats, impacts and the conservation strategies in the African Great Lakes - A review. In *The Limnology, Climatology and Paloeclimatology of the East African Lakes*. Johnson, T.C. and Odada, E.O. (eds). Amsterdam: Gordon Breach, pp. 575-624.

Craig, J. F. The biology of perch and related fish. 1987. Portland, Oregon, Timber Press.

Crul, R.C.M. (1998) *Management and conservation of the African Great Lakes*. Paris: UNESCO.

Cózar, A., Bracchini, L., Dattilo, A., Loiselle, S. and Azza, N. (2004) Characterization of the Ugandan inshore waters of Lake Victoria based on temperature-conductivity diagrams. *Water Resources Research*, 40: W12303, doi:10.1029/2004WR003128.

Denny, P. (1985a) The structure and functioning of African euhydrophyte communities: The floating-leaved and submerged vegetation. In *The ecology and management of African wetland vegetation*. Denny, P. (ed). Dordrecht: Dr. W. Junk Publishers, pp. 125-152.

Denny, P. (1985b) Submerged and floating-leaved aquatic macrophytes (euhydrophytes). In *The ecology and management of African wetland vegetation*. Denny, P. (ed). Dordrecht: Dr. W. Junk Publishers, pp. 19-42.

Denny, P. (1985c) Wetland vegetation and associated life forms. In *The ecology and management of African wetland vegetation*. Denny, P. (ed). Dordrecht: Dr. W. Junk Publishers, pp. 1-18.

Downing, K.M. and Merkens, J.C. (1955) The influence of dissolved oxygen concentrations on the toxicity of un-ionised ammonia to rainbow trout (*Salmo gairdnerii* Rich.). *Annals of Applied Biology* 43: 243-246.

Eggeling, W.J. (1935) The vegetation of Namanve swamp, Uganda. *Journal of Ecology* 23: 422-435.

Emerson, K., Russo, R.C., Lund, R.E., and Thurston, R.V. (1975) Aqueous ammonia equilibrium calculations: effect of pH and temperature. *Journal of the Fisheries Research Board of Canada* 32: 2379-2383.

Gardener, W.S., Cavaletto, J.F., Bootsma, H.A., Lavrentyev, P.J., and Troncone, F. (1998) Nitrogen cycling rates and light effects in tropical Lake Maracaibo, Venezuela. *Limnology and Oceanography* 43: 1814-1825.

Greenwood, P. H. The Cichlid Fishes of Lake Victoria, East Africa. 1974. London, British Museum of Natural History .

Greenwood, P. H. The Haplochromine Fishes of the East African Lakes. Collected papers on their taxonomy, biology and evolution. 1981. München, Kraus International Publications.

Hamilton, A. C. Deforestation in Uganda. 1984. Nairobi, Oxford University Press.

Harper, D.M. (1992) *Eutrophication of freshwaters: principles, problems and restoration*. New York: Chapman & Hall.

Hecky, R.E. (1993) The eutrophication of Lake Victoria. *Verh. Internat. Verein. Limnol.* 25: 39-48.

Holtzman, J. and Lehman, J.T. (1998) Role of apatite weathering in the eutrophication of Lake Victoria. In *Environmental Change and Response in East African Lakes*. Lehman, J.T. (ed). Dordrecht: Kluwer Academic Publishers, pp. pp 89-98.

Jones, R.I. (1992) The influence of humic substances on lacustrine planktonic food chains. *Hydrobiologia* 229: 73-91.

Jørgensen, N.O.G., Tranvik, L., Edling, H., Granéli, W., and Lindell, M. (1998) Effects of sunlight on occurrence and bacterial turnover of specific carbon and nitrogen compounds in lake water. *FEMS Microbiology Ecology* 25: 217-227.

Jude, D.J. and Pappas, J. (1992) Fish utilisation of Great Lakes coastal wetlands. *Journal of Great Lakes Research* 18(4): 651-672.

Kaufman, L. and Ochumba, P. (1993) Evolutionary and conservation biology of cichlid fishes as revealed by faunal remnants in northern Lake Victoria. *Conservation Biology* 7(3): 719-730.

Kirk, J. T. O. Light and photosynthesis in aquatic ecosystems. 1994. Cambridge, Cambridge University Press.

Laegdsgaard, P. and Craig, J. (2001) Why do juvenile fish utilise mangrove habitats? *Journal of Experimental Marine Biology and Ecology* 257(2): 229-253.

Lehman, J.T. and Brandstrator, D.K. (1993) Effects of nutrients and grazing on phytoplankton of Lake Victoria. *Verh. Internat. Verein. Limnol.* 25: 850-855.

Lind, E. M. and Morrison, E. S. East African vegetation. 1974. London, Longman Group.

Lind, E.M. and Visser, S.A. (1962) A study of a swamp at the northern end of Lake Victoria. *Journal of Ecology* 50: 599-613.

Lung'ayia, H., Sitoki, L., and Kenyanya, M. (2001) The nutrient enrichment of Lake Victoria (Kenyan waters). *Hydrobiologia* 458: 75-82.

LVEMP (Lake Victoria Environmental Management Project). The Lake Victoria Integrated Water Quality and Limnology Study. 2002. Dar-es-Salaam, Tanzania, LVEMP Regional Secretariat.

Mazzuoli, S., Loiselle, S.A., Hull, V., Bracchini, L., and Rossi, C. (2004) The analysis of the seasonal, spatial and compositional distribution of humic substances in a subtropical shallow lake. *Acta Hydrochimica et Hydrobiologica* 31(6): 461-468.

Mnaya, B. and Wolanski, E. (2002) Water circulation and fish larvae recruitment in papyrus wetlands, Rubondo Island, Lake Victoria. *Wetlands Ecology and Management*: 133-143.

Morris, D.P., Zagarese, H., Williamson, C.E., Balseiro, E.G., Hargreaves, B.R., Modenutti, B., Moeller, R., and Queimalinos, C. (1995) The attenuation of solar UV radiation in lakes and the role of dissolved organic carbon. *Limnology and Oceanography* 40(8): 1381-1391.

Moss, Brian. Ecology of Fresh Waters: Man and Medium. 1988. Oxford, Blackwell Scientific Publications.

Mugidde, R. (1993) The increase in phytoplankton primary productivity and biomass in Lake Victoria (Uganda). *Verh. Internat. Verein. Limnol.* 25: 846-849.

Mwanja, W.W., Armoudlian, A.S., Wandera, S.B., Kaufman, L., Wu, L., Booton, G.C., and Fuerst, P.A. (2001) The bounty of minor lakes: the role of small satellite water bodies in evolution and conservation of fishes in the Lake Victoria region, East Africa. *Hydrobiologia* 458: 55-62.

Ogutu-Ohwayo, R. (1990) The decline of the native fishes of Lake Victoria and Kyoga (East Africa) and the impact of introduced species, especially the Nile perch, *Lates niloticus*, and the Nile tilapia, *Oreochromis niloticus*. *Environmental Biology of Fishes* 27: 81-86.

Pápistra, E., Acs, E., and Boddi, B. (2002) Chlorophyll-a determination with ethanol - a critical test. *Hydrobiologia* 485(1-3): 191-198.

Randall, D.J. and Tsui, T.K.N. (2002) Ammonia toxicity in fish. *Marine Pollution Bulletin* 45: 17-23.

Scheffer, M., Hosper, S.H., Meijer, M.L., Moss, B., and Jeppesen E (1993) Alternative equilibria in shallow lakes. *Trends in Ecology and Evolution* 8: 275-279.

Scheffer, M. and van Nes Egbert H. (2004) Mechanisms for marine regime shifts: can we use lakes as microcosms for oceans? *Progress in Oceanography* 60: 303-319.

Scheffer, Martin. Ecology of Shallow Lakes. 1998. London, Chapman & Hall.

Schofield, P.J. and Chapman, L.J. (2000) Hypoxia tolerance of introduced Nile perch: implications for survival of indigenous fishes in the Lake Victoria basin. *African Zoology* 35: 35-42.

Scully, N.M. and Lean, D.R.S. (1994) The attenuation of ultraviolet radiation in temperate lakes. *Arch. Hydrobiol. Beih. Ergebn. Limnol.* 43: 135-144.

Seehausen, O. (1996) *Lake Victoria Rock Cichlids: Taxonomy, Ecology and Distribution*. Zevenhuizen, The Netherlands: Verduijn Cichlids.

Seehausen, O., van Alphen, J.J.M., and Witte, F. (1997) Cichlid fish diversity threatened by eutrophication that curbs sexual selection. *Science* 277: 1808-1811.

Seehausen, O., Witte, F., Katunzi, E.F., Smits, J., and Bouton, N. (1997) Patterns of remnant cichlids in southern Lake Victoria. *Conservation Biology* 11(4): 890-904.

Seitzinger, S.P. and Sanders, R.W. (1997) Contribution of dissolved organic nitrogen from rivers to estuarine eutrophication. *Marine Ecology Progress Series* 159: 1-12.

Spence, D.H.N. (1982) The zonation of plants in freshwater lakes. *Advances in Ecological Research* 12: 37-125.

Stephenson, T.D. (1990) Fish reproductive utilisation of coastal marshes of Lake Ontario near Toronto. *Journal of Great Lakes Research* 16: 71-81.

Talling, J.F. (1965) The photosynthetic activity of phytoplankton in East African Lakes. *Int. Rev. ges. Hydrobiologia* 50: 1-32.

Talling, J.F. (1966) The annual cycle of stratification and phytoplankton growth in Lake Victoria (East Africa). *Internationale Revue der gesamten Hydrobiologie* 51(4): 545-621.

Talling, J.F. and Lemoalle, J. (1998) *Ecological dynamics of tropical inland waters*. Cambridge: Cambridge University Press.

Taylor, R. and Howard, K. (2000) A tectono-geomorphic model of the hydrogeology of deeply weathered cystalline rock: evidence from Uganda. *Hydrogeology Journal*, 8: 279-294.

Turner, M.G., Romme, W.H., Gardner, R.H., O'Neill, R.V., and Kratz, T.K. (1993) A revised concept of landscape equilibrium: disturbance and stability on scaled landscapes. *Landscape Ecology* 8(3): 213-227.

Witte, F., Goldschmidt, T., Wanink, J., van Oijen, M., Goudswaard, K., Witte-Maas, E., and Bouton, N. (1992) The destruction of an endemic species flock: quantitative data on the decline of the haplochromine cichlids of Lake Victoria. *Environmental Biology of Fishes* 34: 1-28.

Witte, F., Msuku, B.S., Wanink, J.H., Seehausen, O., Katunzi, E.F.B., Goudswaard, P.C., and Goldschmidt, T. (2000) Recovery of cichlid species in Lake Victoria: an examination of factors leading to differential extinction. *Reviews in Fish Biology and Fisheries* 10: 233-241.

Zagerese, H.E., Diaz, M., Pedrozo, F., Ferraro M., Cravero, W., and Tartarotti, B. (2001) Photobleaching of natural organic matter exposed to fluctuating levels of solar radiation. *Journal of Photochemistry and Photobiology B* 61(1): 35-45.

Chapter 5

Pattern and mechanisms of sediment distribution in Lake Victoria

Submission to a scientific journal based on this chapter:

Nicholas Azza, Johan van de Koppel and Patrick Denny. Spatial distribution and potential dispersal mechanisms of surficial sediments in Lake Victoria (East Africa), *Water Resources Research,* submitted on October 12, 2006.

Pattern and mechanisms of sediment distribution in Lake Victoria

Abstract

An investigation was conducted on the applicability to Lake Victoria of three generalizations on lake sediment distribution namely, that (1) sediment resuspension results almost entirely from the action of wind-induced surface waves; (2) sediments are distributed in a focused pattern; and (3) surficial sediment characteristics (particle size, bulk density, water content and organic matter content) vary systematically with increasing water depth and fetch. From application of two complementary techniques, it was found that the distribution of sediments does not closely follow any of the above generalizations. A belt of proximal-distal progression in sediment character exists in northern Lake Victoria that cannot be explained by wave theory. Surficial sediment characteristics also do not change systematically with depth or fetch. The findings indicated that focusing results only for the case where the wind vector is of variable direction. The anomalies together were taken to suggest a strong influence of currents in sediment distribution. It was accordingly argued that for large lakes with relatively shallow basins, prediction of sediment resuspension and transport from surface wave action without considering the possible influence of currents may not provide a full picture of sediment distribution processes in operation. In processes suggested to explain the anomalies in Lake Victoria, it is proposed that morphometric, hydrological and meteorological factors come together to mix and disperse sediments via the epilimnion, and produce alongshore and cross-isobath currents that transport sediment northwards and northeastwards from the western shore.

Keywords: Wind-induced resuspension, sediment distribution, wave hindcasting, topographic/bathymetric interactions, currents, great lakes, Lake Victoria.

Introduction

Sediments can have an important role in the ecological dynamics of lakes as virtually all types of heavy metals, organic micropollutants and plant nutrients have an affinity for sediments, particularly the fine-grained materials. In large lakes, the resuspension of bottom sediments with adsorbed substances often introduces fluxes of contaminants and nutrients to the overlying water that are much larger than fluxes from external sources (Eadie and Robbins, 1987). To be able to predict water quality problems arising from bentho-pelagic coupling and identify appropriate mitigation measures for them, a good understanding is required, amongst others, of the spatial distribution of sedimentary environments, and the dominant processes controlling sediment resuspension and transport (Håkanson and Jansson, 1983; Harper, 1992). As sedimentary processes lead to formation of sedimentary deposits, an improvement in the understanding of modern sediment distribution mechanisms, besides aiding lake management, could lead to improvements in interpretation of sedimentary records in paleoclimatic and peleolimnological studies.

The distribution of sediments in lakes is a complex process resulting from the interaction between various confining factors such as bathymetry, size, and shape of basin; forcing functions such as wind and currents; and the differential response of particles of various sizes and suspended load concentrations (Sly, 1978; Evans, 1994). The physical agents that resuspend and transport sediments are numerous and include wind, fluvial currents, slow-moving lacustrine circulations, wind-, wave- and storm-induced currents and density flows (Reading and Levell, 1996).

In lakes it is generally considered that nearly all of the sediment resuspension is due to the action of wind-induced surface waves (Håkanson, 1977; Håkanson and

Jansson, 1983; Rowan *et al.*, 1992; Evans, 1994) though at any point in time dynamic processes driven by one or several of the above physical agents will be in operation. Current resuspension occurs in lakes but in many cases current velocities have been noted to be too weak to stir up significant quantities of bottom sediments into flow (Hilton *et al.*, 1986; Bengtsson *et al.*, 1990; Leuttich *et al.*, 1990; Hawley and Lesht, 1992). Similarly, river delta formation and river plume sedimentation are considered to only make minor contributions to sediment distribution except in cases where there is very high loading of terrigenous material (Håkanson and Jansson, 1983; Hilton *et al.*, 1986). Given these past experiences, there has been a tendency for new investigations on sediment distribution in lakes to ignore the possible role of alternative sediment resuspension processes besides wind-induced resuspension without first carry out some kind of scoping.

Wind acting on the surface of a lake produces progressive surface waves at the lake surface, and orbital molecular movements in the water column, that are attenuated with increasing depth. In areas where the lake is shallow enough for the wave base to 'touch' the bottom, bed scour and sediment resuspension take place (Pond and Pickard, 1983). Large and more dense particles in the suspended sediment mass fall back quickly to the bottom while fine and less dense ones remain in suspension for longer periods and are gradually transported by currents (over the long term) to depths where the wave base no longer disturbs the lakebed.

The above process, termed sediment focusing, leads to the dominance of coarse-grained particles along lakeshores, and fine-grained sediments (silt and clay; < 20 μm) in deep offshore areas (Thomas *et al.*, 1972, 1973, 1976; Evans and Rigler, 1980, 1983; Hilton, 1985). Focusing, furthermore, normally produces systematic changes in sediment characteristics (particle size, bulk density, water content and organic matter content) with increasing water depth and fetch (Håkanson and Jansson, 1983; Rowan *et al.*, 1992; Blais and Kalff, 1995). Fetch is a measure of the free water surface over which the transfer of wind energy takes place. Another morphometric parameter, slope, often confounds the sorting effect of wind-wave resuspension and transport. When greater than 4%, basin slope causes sliding and slumping of deposited sediments (Håkanson, 1977; Rowan *et al.*, 1992; Blais and Kalff, 1995).

Sediment resuspension and transport in lakes has mainly been noted to occur at times of episodic events, during which extreme wind energies are transferred to the whole lake (Evans, 1994; Eadie *et al.*, 1996). Rowan *et al.* (1992), from analysis of wave height exceedance data for lakes Superior and Huron, concluded that the one or two biggest storm events each year are the ones responsible for sediment distribution. Similarly, Lick *et al.* (1994), using modeling tools and geochronological data showed, for Lake Erie, that major storms, despite their infrequent occurrence, cause the most transport of any class of wind events, and are responsible for more of the resuspension and transport of sediments than the total of all the lesser storms and wind events.

There are upper limits to the size of particles that are resuspended and transported by the different physical agents: only particles finer than 100 μm (very fine sand) are normally picked up in suspension; particles between 100-1000 μm are readily moved as bedload; and particles larger than 1000 μm (very coarse sand) are only moved by exceptionally strong winds as bedload (Reading and Levell, 1996).

These size constraints give rise to three distinct sediment populations: clay and silt (fine-grained sediments) transported in suspension; sand transported as bedload; and gravel that is seldom moved.

Håkanson (1977) introduced a classification for lake bottoms that is widely used. On the basis of the likelihood of deposition of fine-grained sediments, he divided lake bottoms into three: erosional zones, where resuspension occurs continuously and there is no deposition of fine-grained particles; transportation zones, where resuspension occurs intermittently and fine-grained particles deposit temporarily; and accumulation zones, where fine-grained materials deposit continuously and permanently. Sediments of a particular type characterize each zone: heavy, coarse-grained sand and gravel characterize the erosional zone; light, fine-grained silt and clay characterize the accumulation zone; and materials of intermediate texture characterize the transportation zone.

Much of the current knowledge on lacustrine sediment distribution has come from studies in small lakes. The main mechanisms by which sediment is distributed in large lakes are not well understood. In terms of size, large lakes lie between small lakes and oceans, and compared to either of these, they have been studied less intensively, tropical large lakes even more so (Johnson, 1984). Hilton *et al.*, (1986) from reviewing the literature on small lakes identified five dominant mechanisms by which wind and wind-induced surface waves effect sediment distribution. These they enumerated as (a) continuous complete mixing (of the whole lake); (b) intermittent complete mixing; (c) intermittent epilimnetic complete mixing; (d) peripheral wave attack; and (e) random redistribution of sediments.

No similar list has been compiled for large lakes and, although all of the above mechanisms may be expected to be in operation in large lakes, there are probably differences regarding the relative importance of each. Hilton (1985), for example, argued that complete mixing is unlikely to dominate other mechanisms in large lakes because, compared to small lakes, the thermocline in large lakes is destroyed more slowly, and the high water velocities needed to resuspend particles from deep water and mix them efficiently into the water column are less likely to be maintained for the extended overturn period. In large lakes, furthermore, the likelihood for strong influence of currents in sediment transport is greater than in small lakes.

The objectives of this study are two. The first is to assess how well present generalizations on sediment resuspension and transport, which mainly come from small lakes, explain the spatial variation of sedimentological zones and surficial sediment characteristics in Lake Victoria (a lake with a large open surface but relatively shallow basin). In particular, the applicability of three common assumptions is tested, namely that: (a) wave-induced resuspension and slope are responsible for much of the distribution pattern of sediments in the lake; current resuspension need not be considered; (b) sediments are distributed in a focused pattern with sand and gravel occurring around basin rims and silt and clay occurring in the deep offshore regions; and (c) sediment characteristics (particle size, bulk density, water content and organic matter content) change systematically with increasing water depth and fetch. The second objective of the study was to determine the pathways of sediment transport in the lake and processes responsible for producing them.

Materials and Methods

Study area

Lake Victoria, found in East Africa, is the world's second largest freshwater lake ($68{,}870$ km^2, 40 m mean depth, 82 m maximum depth). Large-scale winds over the basin are mostly associated with the monsoon phenomenon in the Indian Ocean. The predominantly easterly air currents from the Indian Ocean are modified seasonally into SE and NE trade winds by the position and annual migration of the Inter-Tropical Convergence Zone (ITCZ), and the positions of adjacent sub-tropical high-pressure systems (Griffiths, 1972; Nicholson, 1996). The NE trade winds blow between December and February and the SE trade winds between June and September. The region further receives equatorial westerlies from the Atlantic Ocean and the moist Congo basin in July and August that are induced by the maximum northward shift of the ITCZ (Nicholson, 1996).

Within the lake are several archipelagos that shield parts of the shoreline from wind action and introduce heterogeneity in the spatial distribution of wind and wave exposure. Superimposed on the annual cycle of prevailing winds is a marginal diurnal cycle of land and lake breezes. Rainfall is bimodal, with long rains falling in March-June and short rains in October-December. Several rivers drain into the lake, the largest of which, the Kagera River, delivers a considerable load of sediments on the western coast. Offshore thermal stratification occurs in three phases with between-year variation in the start and end of the cycle. The warming phase is from December to April and is characterized by progressive downward displacement of the primary metalimnion with attendant warming of the lake. The start of the cooling phase coincides with the onset of SE trade winds in late May to June and culminates in the annual overturn of the lake. The final phase is from September to November and is characterized by destratified or weakly stratified conditions (Talling, 1966; LVEMP, 2002).

Overview of approach

Two standard approaches were combined in the study. The two – prediction of the accumulation zone boundary and surficial sediment pattern analysis – give different but complementary information. First, wind pattern analysis was carried out to generate information on the direction, magnitude, duration, frequency and annual variation of winds. Then, the wind data was used to make predictions of the position of the transportation/accumulation boundary based on wave theory, and compared the predicted patterns with the patterns obtained from field sampling/measurement of surficial sediment characteristics. Good agreement between the predicted and observed transportation/accumulation boundaries was taken to confirm the dominance of wind-induced surface wave resuspension. On the other hand, lack of agreement, or large areas of disparity, was taken to suggest other resuspension mechanisms to be dominant, or at least important.

Transport pathways were deduced from studying the spatial pattern of surficial sediment characteristics. Several previous studies have successfully made deductions on transport pathways and sediment distribution mechanisms from analyzing the spatial variation of grain size and related physico-chemical characteristics of surficial sediments (Folk and Wand, 1957; McLaren, 1981; Cheng *et al.*, 2004). This is possible because some of the commonly measured physico-chemical parameters like grain size and bulk density are among the critical physical properties determining the relative ease with which sediments are entrained, transported or settled out of suspension (Håkanson and Jansson, 1983; Jepsen *et al.*, 1997). Other parameters like water content and organic matter content are strongly correlated to the basic physical properties (Thomas, 1969; Håkanson, 1977; Blais and Kalff, 1995).

Finally, proposals have been made on potential sediment distribution processes from a consideration of the wind regime, basin morphometry, annual cycle of stratification and spatial variation of surficial sediment characteristics.

Analyzing wind patterns
Ten years (1995 to 2004) of wind data recorded at Entebbe International Airport (WMO Centre 750, Station No. 89320660) located on the northern shores of Lake Victoria were acquired. Earlier studies (Newell, 1960) showed weather conditions to vary systematically from south to north in the lake, but noted that the same diurnal and seasonal trends occur across the basin. The Entebbe data can hence be taken to be indicative of general conditions over the whole basin. The data, provided by the Department of Meteorology, Uganda is based on minute averages of wind force (in knots) and direction (in degrees) taken on the hour. There were 74,071 records excluding missing data. Checks on the data by comparison with direct measurements on the lake found good correlation.

After quality checks, the data was processed to obtain frequency distributions for the whole data period. The frequency distributions were for instantaneous wind speeds and directions (for 16 compass directions) at different times of day and different times of year. Winds with speeds equal to or greater than 12 knots (6 m s^{-1}; 21.6 km h^{-1}) and 17 knots (8.75 m s^{-1}; 31.5 km h^{-1}) were also selected to examine the daily and annual pattern of strong winds. Selection of the stronger winds allows the characterization of large-scale air-mass flows.

For winds to produce an effect in lake mixing and sediment resuspension, they must be sufficiently strong and persist for several hours. Carper and Bachmann (1984) have suggested that winds must be in excess of 15-20 km h^{-1} (4-5.5 m s^{-1}; 8-10.8 knots) to produce resuspension in most systems. A minimum wind duration of three hours was selected that, from consideration of the long fetches over the lake, should be sufficient to produce fully developed conditions in the lake for high wind speeds. Using three-hourly-averaged conditions from a 3-point moving average, frequency distributions of sustained wind speeds and directions were derived. Applying non-linear regression techniques to the frequency distribution of sustained speeds, a value of 10 m s^{-1} (19.4 knots; 36 km h^{-1}) was obtained as the sustained wind speed that is exceeded only two times in a year (occurs 0.023% of the time). This value, used in the wave resuspension calculations, was assumed to represent the force of episodic strong winds occurring infrequently but often enough to be significant on a geomorphic scale.

Predicting the position of the boundary between transportation and accumulation zones from wave theory

An approach based on the Airy deepwater wave theory and particle threshold theory was used in which a prediction of whether or not sediment resuspension occurs at a given point in a lake is made by comparing a computed value of maximum horizontal velocity U_m (m s^{-1}) of deepwater waves at or near the lakebed, with an estimate of the critical horizontal velocity U_{cr} (m s^{-1}) for the onset of erosion of fine-grained sediments (Bengtsson *et al.*, 1990; Weyhenmeyer *et al.*, 1997). The maximum horizontal velocity U_m was estimated from the expression (Sheng and Lick, 1979)

$$U_m = \frac{\pi H}{T \sinh\left(\frac{2\pi h}{L}\right)} \qquad (5.1)$$

where H (m) is deepwater wave height, T (s) is deepwater period, L (m) is deepwater wavelength and h (m) is water depth at the measurement point. If it can be assumed that wind duration is not limiting, then significant wave height, wave period and wavelength required in equation (5.1) to reflect maximum wave conditions, are given by the empirical models developed by the Beach Erosion Board (CERC, 1984)

$$\frac{gH}{w^2} = 0.0026\left(\frac{gF}{w^2}\right)^{0.47} \qquad (5.2)$$

$$\frac{gT}{w} = 0.46\left(\frac{gF}{w^2}\right)^{0.28} \qquad (5.3)$$

$$L = 20H \qquad (5.4)$$

where g (m s^{-2}) is acceleration due to gravity, w (m s^{-1}) is wind speed and F (m) is effective fetch. The term 'significant' above refers to the mean of the largest one third of values. For a total of 935 points spaced regularly (on a 10 x 10 km grid over open water) and irregularly (between and around islands; the inner parts of small bays), the outcome of $U_m - U_{cr}$ was calculated and the boundary of the accumulation zone (i.e. the isoline for $U_m - U_{cr} = 0$) obtained by exact interpolation of $U_m - U_{cr}$ using the Natural Neighbour gridding technique.

Water depth, h, required in the calculations was obtained from a bathymetric map created by digitising British Admiralty Charts No. 3252 and 3665 of The Victoria Nyanza (Figure 5.1). For wind speed w, a value of 10 m s^{-1} (see preceding section) was used. The concept of effective fetch F was developed to take into account the decrease in effectiveness of wind energy transfer as one moves away from the mean wind direction. Effective fetch F was calculated at each of the 935 points from a 1:250,000 map sheet using the expression (Håkanson and Jansson, 1983; CERC, 1984)

$$F = \frac{\sum x_i \cos a_i}{\sum \cos a_i} \qquad (5)$$

where x_i (m) is the fetch length or straight-line distance from the point of fetch measurement to land or an island and a_i (degrees) is the angle from the wind direction azimuth in $6°$ increments from $+42°$ to $-42°$. The final value required for the computations was that of U_{cr}. Several determinations have been made of the critical horizontal velocity of water needed to resuspend freshly deposited fine-grained materials. Reported values in literature range from 0.005 to 0.017 m s^{-1} (Sheng and Lick, 1979; Gardner et al., 1983; Bengtsson et al., 1990). A value of 0.01 m s^{-1} for the critical horizontal velocity was used in this study.

With the above procedure, the pattern of fine-grained sediment deposition was predicted for four wind directions: north, northeast, southeast and south. These are the dominant directions of strong winds obtained from the wind data (discussed below). In a fifth case, it was assumed that wind at each point in the lake comes from the direction with longest fetch. This case represents a situation where the wind vector direction varies from site to site and from the prevailing wind direction, which can happen during episodic violent storms. For the fifth case, maximum effective fetch MEF replaces effective fetch F in equations (2) and (3). MEF is the outcome of equation (5) when the wind direction azimuth is oriented in such a way as to give the longest possible straight-line distance from a measurement point to land or island.

Determining the spatial variation of surficial sediment characteristics in northern Lake Victoria

Due to the great size of the lake, fieldwork was limited to its northern half (Figure 5.1). Eighty-five samples were collected between 2002 and 2004 from an area measuring roughly 35,500 km^2 (approximately one sample per 20 x 20 km area). Fine muds were sampled with a small gravity corer (Technical Ops Corer) and the upper 0-3 cm layer taken for analysis. Coarse unconsolidated sands were sampled with a lightweight Ponar grab sampler. The samples were analysed for water content, organic matter content (as loss-on-ignition), wet bulk density and grain size distribution in the Water Quality Laboratory of the Department of Water Resources Management in Entebbe, Uganda.

Water content and loss-on-ignition were determined by drying samples to constant weight at 105 °C and igniting to constant weight at 550 °C respectively (APHA, 1995) while wet bulk density was determined with a pycnometer (Guy, 1994). Grain size distribution in the samples was determined by sieving and hydrometer test after removing organic matter and carbonates with hydrogen peroxide and hydrochloric acid respectively (ASTM, 1986). The grain size distribution results were used to determine the sand content (diameter 63 – 2000 μm) and mud content (diameter <16 μm) of samples, and to compute the median (M_d), sorting (S_o), and skewness (α_s) of the distributions from the expressions (Håkanson and Jansson, 1983):

$$M_d = \phi_{50} \quad (6); \qquad S_o = \frac{(\phi_{95} - \phi_5)}{2} \quad (7); \qquad \alpha_s = (\phi_{95} + \phi_5) - 2\phi_{50} \quad (8)$$

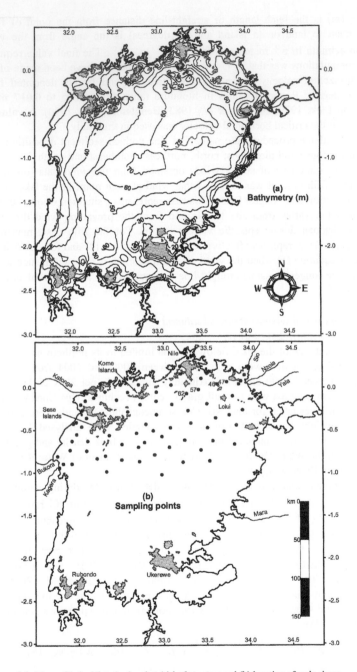

Figure 5.1: Maps of Lake Victoria showing (a) bathymetry; and (b) location of main rivers, islands and sediment sampling points. The islands are the grey-filled polygons. The numbered sites in Figure 5.1(b) are discussed in the text in relation to anomalies in sediment characteristics. Map coordinates are in decimal degrees.

The median was chosen over the mean as it provides a better measure of central tendency in strongly skewed distributions (Håkanson and Jansson, 1983), which is the case with Lake Victoria sediments. Sorting and skewness are valuable tools for investigating the genealogy of sediments (Folk and Wand, 1957; Visher, 1969; McLaren, 1981).

The spatial variation of each sediment parameter was obtained by gridding and exact interpolation of parameter values at the 85 sampling points using Kriging gridding method. Water depth, slope and fetch were also determined at each sampling point. Depth was measured in the field with an echo sounder, slope was calculated from the digital bathymetric map and fetch was calculated with equation (5) for the five cases discussed above.

Results

Wind regime
The pattern of air-mass circulations that was found agrees closely with patterns established by an earlier study (Flohn and Fraedrich, 1966). Airflows over Lake Victoria closely follow the regional wind regime and consist mainly of flows of weak magnitude and multiple directions (Figure 5.2 and 5.3). SE and NE trade winds that dominate regional large-scale air-mass circulations show up over the lake as strong northerly and southerly wind currents. Northerlies dominate flows between November and February while southerlies and southeasterlies dominate flows between May and October. Periods around the change from one wind system to the other (March-April and September) have roughly equal frequencies of northerlies and southerlies. There are occasional cases of westerlies, especially between February-May and in September.

Calms on the lake occur for only 3.5% of the time while light airs and light breezes (1-6 knots; $0.5 - 3$ m s^{-1}) occur for two-thirds of the time. Moderate to strong instantaneous winds (≥ 11.7 knots; 6 m s^{-1}) occur 250 times (out of 8,784 possibilities) in an average year. The highest speed in the data set was a Force 10 storm of 50 knots (25.7 m s^{-1}; 93 km h^{-1}) recorded on 3 May 2003.

The frequency distribution of instantaneous wind speeds over the time of day is unimodal, with the strongest winds, and possibly severest storms, occurring in the night between 2300 and 0300 hrs while the calmest time of day is from early to late afternoon (1300-1900 hrs) (Figure 5.4).

Strong winds, though infrequent, occur at all times of the year with twin peaks in March-May and September-November. Cases when the strong winds are sustained for three or more hours are very few (Figure 5.4). The frequency of the more powerful winds is strongly biased towards northerly and southerly directions although the mean force of strong winds is nearly even for all compass directions (Figure 5.3).

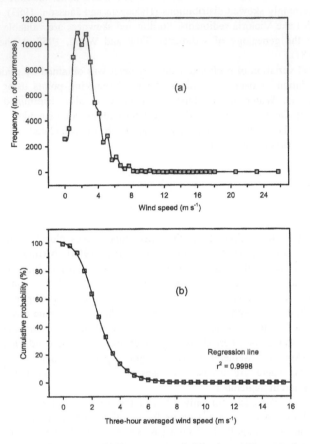

Figure 5.2: Frequency distribution of (a) instantaneous wind speeds; and (b) sustained strong winds (≥ 8.75 m s^{-1}) at Entebbe. Distribution (b) gave a speed of 10 m s^{-1} (36 km h^{-1}; 19.4 knots) as the force of strong sustained winds equaled or exceeded only twice in a year.

Distribution of sedimentary zones predicted from wave theory

The pattern of fine-grained sediment distribution predicted by wave hindcasting techniques for four wind directions (northerly, northeasterly, southeasterly and southerly) and for a fifth case with spatially variable wind direction are shown in Figures 5.5 and 5.6. Generally, the wave hindcasting techniques predicted large areas (about 70%) of the bottom of Lake Victoria to be overlain by fine-grained sediments. There was reasonable agreement between the location and spatial extent of predicted accumulation zones and actual zonation based on sediment sampling thereby suggesting dominance of surface-wave action over other potential resuspension mechanisms. There were also regions of disparity that are discussed more fully below.

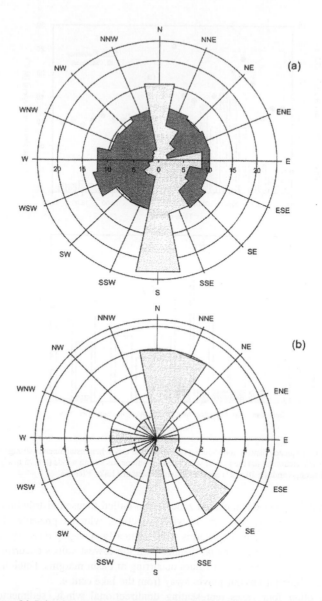

Figure 5.3. Polar plots showing the directional distribution of (a) instantaneous strong winds (≥ 8.75 m s^{-1}); and (b) sustained strong winds (≥ 10 m s^{-1}) at Entebbe, northern Lake Victoria. The polygons in (a) represents mean wind speeds in m s^{-1} (the dark-gray shape) and frequency as number of occurrences (light-gray shape). In (b) the polygon represents frequency (number of occurrences).

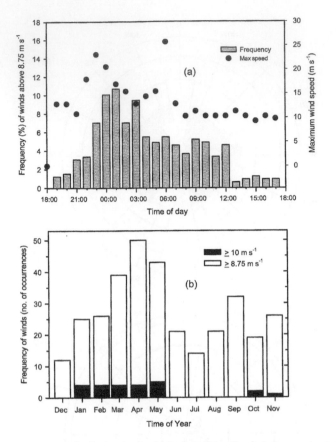

Figure 5.4. Frequency distribution showing (a) the time of day when instantaneous strong winds (≥ 8.75 m s^{-1}) occur; and (d) the time of year when instantaneous strong winds (≥ 8.75 m s^{-1}) and sustained strong winds (≥ 10 m s^{-1}) occur at Entebbe, northern Lake Victoria.

A centrally-located accumulation zone, the so-called focused distribution pattern results only for the case of variable wind direction, which represented the case where sediment is distributed by storms with erratically changing direction. For this case, fetch is distributed in a radial manner with the lowest values occurring at the centre of the lake and the highest values occurring at basin margins. Fetch increases steadily in all directions as one moves away from the lake centre.

For the other four cases representing unidirectional winds, sedimentological zones are distributed along a linear upwind-downwind axis. Effective fetch is lowest along the upwind shore and increases steadily across the lake reaching a maximum at the opposite shore. Fine-grained sediments are distributed in an opposite manner: the boundary between transportation and accumulation zones is shifted towards the upwind shore, away from the region of high fetch and strong bed scour at the opposite downwind shore.

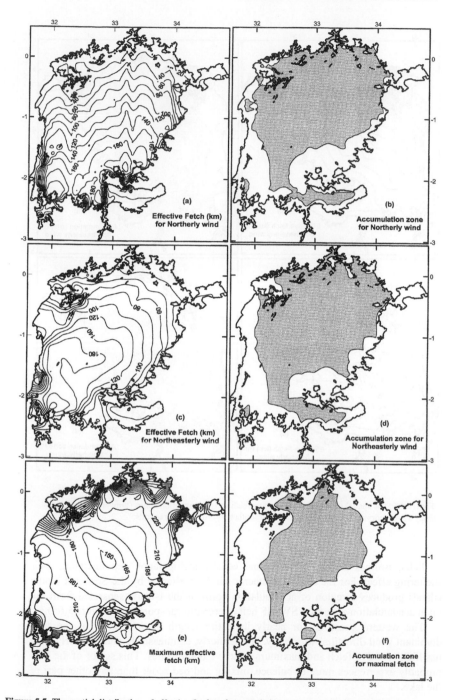

Figure 5.5: The spatial distribution of effective fetch and accumulation zones (gray area) predicted from wave theory for northerly (a+b) and northeasterly (c+d) wind directions, and for wind of variable direction (e+f).

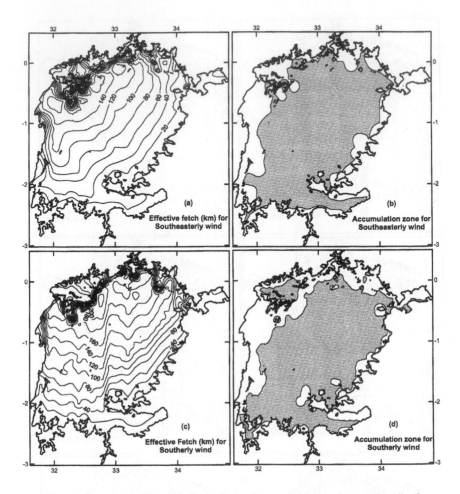

Figure 5.6: The spatial distribution of effective fetch and accumulation zones (gray area) predicted from wave theory for southeasterly (a+b) and southerly (c+d) wind directions.

Fetch isolines are clustered around coastal archipelagos demonstrating their sheltering effect. For the case of southerly winds, sheltering by the Sese and Kome islands produces a region of accumulation north of the islands separated from the main accumulation zone by a SW-NE belt of erosion-transportation (Figure 5.6d).

The western coastline for nearly its entire length is downwind of the four dominant wind directions (northerly, northeasterly, southeasterly and southerly) and has high effective fetch and maximum effective fetch values. This sector of the lake probably experiences more turbulence, mixing and bed scour than any other part of the lake.

Spatial distribution of surficial sediment characteristics in northern Lake Victoria

The textural composition of sediments at the sampled points is shown in Figure 5.7. Most sites had a high content of silt and clay, and low content of sand. The commonest texture types in order of decreasing frequency are silty clay, clay, clayey silt and sand. The spatial pattern of surficial sediment characteristics was anomalous in two respects: (a) sediment characteristics do not vary systematically with water depth or effective fetch (sediment characteristics were correlated with fetch values for the five cases used in the theoretical predictions); and (b) distribution is strongly directional with a distinct linear belt of proximal-distal progression in character. These anomalies are further elaborated below.

Coarse- and fine-grained sediments occur in shallow as well as deep waters, and in sheltered as well as exposed sites. The lack of correlation with depth and fetch shows in all measured sediment parameters (Figure 5.8). Surficial sediments are further distributed in a manner suggesting strong directional forcing. On the western coast is a small sector of erosional character dominated by bulky, coarse-grained sands. From this zone extends a peculiar cone-shaped region of fine sand

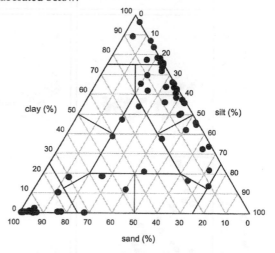

Figure 5.7. A ternary plot showing the textural composition of surficial (0-3 cm) sediments in northern Lake Victoria. The texture classification is according to Shepard (1954).

followed by a broad belt of silt that stretches to the opposite northeastern shore, effectively bisecting the region of clay in the north into two.

The above systematic trend in texture, which seems to describe the trajectory of a major turbidity plume, is also evident in the spatial distribution of the other sediment parameters (Figures 5.9 and 5.10). The percentages of sand and mud have opposite gradients along this pathway, suggesting progressive waning in the force causing this remarkable distribution. The phenomenon produces two regions within the accumulation zone in the northern half of the lake that are characterized by relatively low bulk density (< 1.15 g cm^{-3} ws), low sand content ($< 10\%$), high mud content ($> 80\%$), high organic matter ($> 25 \%$ as LOI) and high water content ($> 90\%$). One region lies on the northern peripheries while the other stretches from the lake centre to the northeastern margin. The SW-NE belt of heterogeneity and proximal-distal differentiation separates the two regions. The spatial pattern of most surficial sediment characteristics shows this belt to curve southwards near the lake centre.

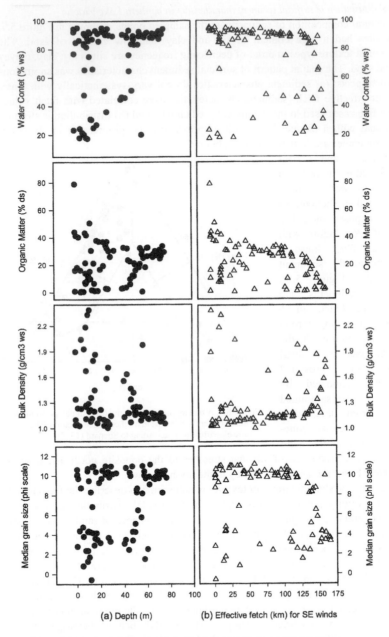

Figure 5.8. Scatter diagrams depicting relationships between surficial sediment characteristics and (a) water depth; and (b) effective fetch (for the case of SE wind). Sediment characteristics do not correlate with the two morphometry variables.

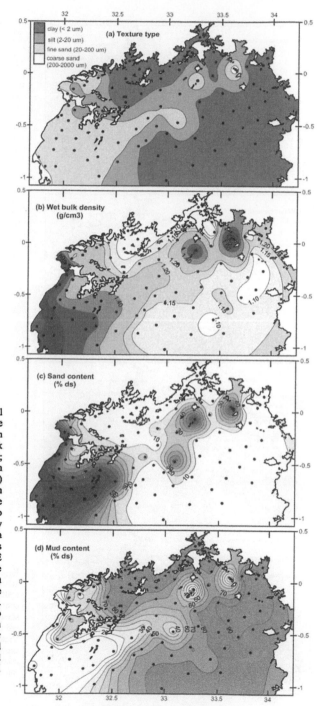

Figure 5.9. The spatial distribution of (a) texture types based on median grain size; (b) wet bulk density; (c) sand content; and (d) mud content in surficial (0-3 cm) sediments of northern Lake Victoria. The distribution depicts two regions of fine clayey sediments bisected by a belt of silty materials stretching along a SW-NE axis. Texture classes are according to Håkanson (1977). The dots show the location of sampling sites. Shading is used to highlight regions with fine-grained particles, low bulk density, low sand content and high mud content, which are typical of accumulation zones.

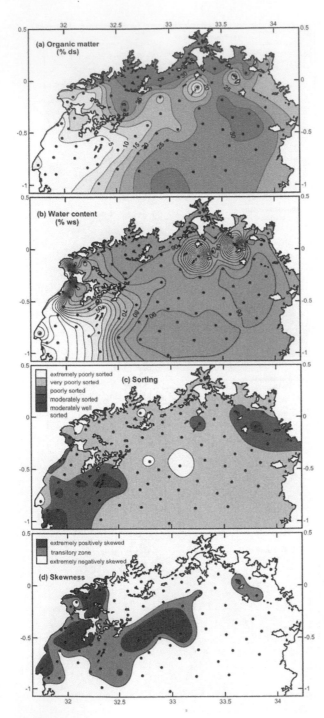

Figure 5.10. The spatial distribution of (a) organic matter (as LOI); (b) water content; (c) sorting; and (d) skewness in surficial sediments of northern Lake Victoria. Sorting and skewness classes are according to Håkanson and Jansson (1983).

Slope is low all over northern Lake Victoria. At the sampled sites, it ranged from 0.0001 to 0.357%. At such low values, it is considered not to influence sediment distribution (Rowan *et al.*, 1992). Like water depth and fetch, slope was not correlated with any surficial sediment property. It was also low along the SW-NE belt of systematic change ($\underline{M}$ = 0.019%; $\underline{SD}$ = 0.066%) hence making it unlikely that this belt defines the path of a gravity flow.

Sorting of sediments is generally very poor in the northern part of the lake, with the regions on the northwestern and northeastern shores being relatively better sorted than the rest of the areas. The relatively better sorted areas are entry points of major rivers (see Figure 5.1). Sorting, which is a measure of the spread of particle sizes around the mean, provides an indication of the influence of flow on sediment deposition. Well-sorted sediments in a site point to a frequent or continuous occurrence of hydraulic flow condition at the site. Conversely, poorly sorted sediments point to rare or intermittent occurrence of hydraulic flow (Visher, 1969; Håkanson and Jansson, 1983). The results of this study hence suggest relatively regular flow to occur in the areas adjacent to the mouth of the Rivers Kagera, Nzoia and Yala, and rarely in the rest of the lake.

The frequency distribution of grain size in most parts of northern Lake Victoria is either extremely positively skewed or extremely negatively skewed. Skewness is a measure of symmetry in the distribution of particle sizes. It provides an indication of the extent of particle mixing and magnitude of environmental energy that transports sediments (Folk and Wand, 1957; Damiani and Thomas, 1974). Positive skewness indicates clustering of scores on the left side of a distribution (the coarse-grained end of a phi scale) while negative skewness indicates clustering of scores on the right hand side of a distribution (the fine-grained end of a phi scale). The extreme positive skews in Lake Victoria are caused by the presence of fine sand and silt in predominantly coarse sand while the extreme negative skews are due to very fine sand and silt in predominantly clayey material. The two extremes suggest the occasional movement of sediments by high-energy processes in Lake Victoria. Between the regions of extreme opposite skews is a small zone of transition where the value of the skewness parameter changes rapidly.

There are relationships between the different sediment characteristics that are revealed when distribution maps are compared. From Figures 9 and 10, it can be seen that areas with coarse-grained sediment also have high wet bulk density, high sand content, low mud content, low organic matter and low water content. The opposite is true for sites with fine-grained sediments.

Further linkages are revealed when maps of skewness, sorting and sedimentological zones are overlaid (Figure 5.11). Areas of extremely positively skewed sediments can be seen to lie on the northern and northeastern peripheries of areas of erosion-transportation, which are also areas of better-sorted sediments. This suggests a high frequency of resuspension on the western coast and a northward and northeastward movement of materials resuspended from there. The presence of a small transportation zone and region of extremely positively skewed grain size distribution at the mouth of the Kagera River suggests some transportation to occur there under the influence of the river.

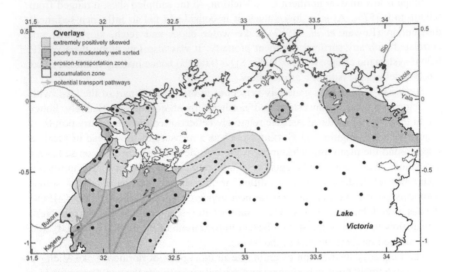

Figure 5.11. Overlays of sedimentological zones with selected classes of grain size sorting and skewness. Areas of sediments with extremely skewed particle size distribution lie to the north and northeast of zones of erosion-transportation, suggesting northward and northeastward transport of sediments from the western part of Lake Victoria.

There are two sites, numbers 46 and 62 that produce "bull's eye" type of concentric circles in the distribution of most parameters. Sampling and analysis was repeated three times at these sites but produced no improvement in values. The anomalies are thus not artefacts of erroneous laboratory analysis but are reflective of large local variability in sediment character as illustrated in Figure 5.12 for site 46.

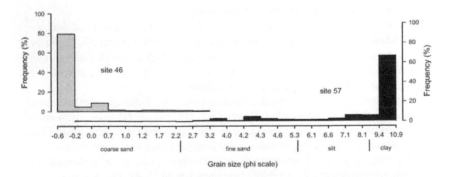

Figure 5.12. Particle size frequency distribution in surficial sediments at sites 46 and 57 showing the contrasting textural composition at the two neighbouring sites. Another neighbouring site, number 47, had a grain size distribution similar to that of site 57.

Seeing a possible important role for River Kagera in the sedimentology of Lake Victoria, mean monthly flows of the river (Figure 5.13) were obtained. The river delivers 4250 x 10^6 m^3 yr^{-1} of water, which is about 30% of the total annual river input to Lake Victoria. The remaining 70% is distributed amongst 23 rivers around the basin (Yin and Nicholson, 1998). The combined flow of the Bukora and Katonga, the other western rivers, is less than 2% of the flow of the Kagera River. The annual peak in flow, when most sediment is expected to be delivered, is between May and August, coinciding with the period of annual overturn and destratification in Lake Victoria. Data on monthly sediment inputs was not readily available but recent measurements by the Directorate of Water Development in Uganda estimate the load at 728 x 10^3 tons yr^{-1} (Mwebembezi, pers. comm.).

To ascertain the mechanism of river plume dispersion, in February 2002 CTD (conductivity, temperature and depth) profiles were measured with an YSI 600 QS multiparameter probe and data sonde (Yellow Springs Inc. Ohio) at the mouth of the Kagera. The acquired longitudinal temperature profile (Figure 5.14a) indicates that the river does not persist as a distinct plume at any depth. At the time of measurement, the river was colder than the lake by about 3 C^o which, in the temperature range of 22-26 °C, should have been sufficient to produce strong density stratification (Denny, 1972; Talbot and Allen, 1996). Relentless wave attack on the western coast may be responsible for the thorough mixing of river and lake waters. On the other hand, the longitudinal profiles of bathymetry and electrical conductivity (Figure 5.14b) suggest that much of the river suspension (probably coarse sand fraction) is deposited at the river mouth while some material is transported into the lake as bedload.

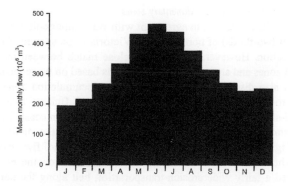

Figure 5.13. Long-term (1963-2000) mean monthly flows of River Kagera at Kyaka Bridge. Data is from the Directorate of Water Development, Uganda

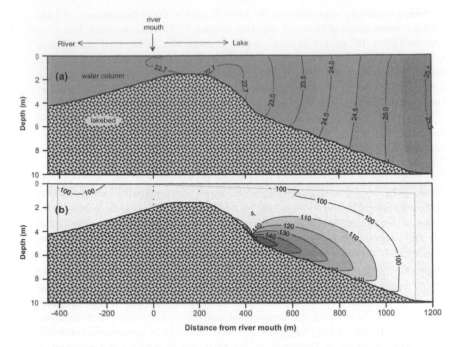

Figure 5.14. Diagrams showing longitudinal depth-distance variation of (a) water temperature (in $^{\circ}$C); and (b) electrical conductivity (in μ S cm^{-1}) at the mouth of River Kagera. Contours were obtained by applying the Natural Neighbor gridding method to vertical CTD profiles measured at 50 m intervals along a linear transect from river to lake.

Comparison of predicted and observed sedimentary zones

Surficial sediment characteristics, in agreement with wave hindcasting results, show large areas (about two-thirds) of northern Lake Victoria to be sites for fine-grained material accumulation. However, there was no close match between the observed sedimentological zones and any of the five predictions based on wave theory (Figure 5.15). The grain size distribution, besides the large accumulation zone, shows a small area of erosion at the mouth of the Kagera and Bukora rivers, and an extensive, funnel-shaped transportation zone on the northwestern coast. There is also a small transportation zone extending southeastwards from the mouth of River Katonga (see Figure 5.1 for location of main rivers). The five cases either underestimated the extent of the erosion-transportation zone on the northwestern coast, or predicted a fairly wide erosion-transportation belt along the northern and northeastern peripheries of the lake whereas that area in reality is an accumulation zone.

There appeared to be co-dominance between northerly and southerly winds. Of the five cases, predictions based on the influence of northerly winds had a better fit for the extensive accumulation zone on the northern and north-eastern margins of the lake, but underestimated the extent of the western erosion-transportation area. On the other hand, the prediction based on the influence of southerly winds was the only one that had a large erosion-transportation zone on the northwestern coast, as well as an erosion-transportation belt wedged between accumulation zones that resembled the belt of proximal-distal differentiation in surficial sediment characteristics.

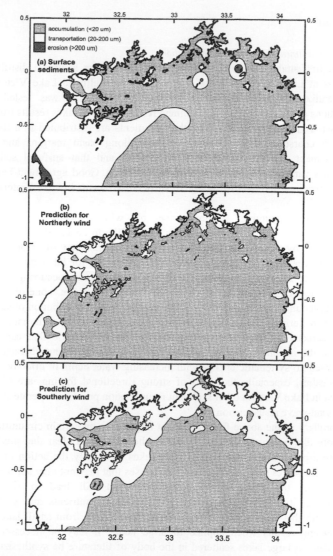

Figure 5.15. Comparison of spatial distribution of sedimentary environments based on (a) surficial sediment characteristics; and in (b) and (c), wave hindcasting techniques. The light-grey areas represent accumulation zones for fine-grained sediments. The unfilled area in (a) represents the transportation zone while in (b) and (c) it represents the combined erosion and transportation zones.

Outside of the regions of disparity, there was a close match between the predicted and actual distribution of sedimentary zones. This suggests that the wind speed of 10 m s^{-1} used in the calculations is a good approximation of the speed of sustained winds responsible for wind-induced sediment resuspension in the lake.

Discussion

Applicability of sediment distribution generalizations
This is the first study to investigate systematically modern sediment distribution mechanisms in Lake Victoria. In the study, the applicability to Lake Victoria of three generalizations on sediment distribution within lakes was tested. The generalizations are that nearly all the sediment resuspension in lakes results from the action of wind-induced surface waves; that sediments are distributed in a focused pattern with coarse-grained sediment occurring along basin margins and fine-grained sediment occurring in deep water zones; and that surficial sediment characteristics vary systematically with depth and fetch. Good agreement between these generalizations and the distribution of surficial sediments in Lake Victoria was not found.

Computations assuming a dominance of wind-induced surface wave resuspension correctly predicted large areas of the lake to be lain by fine-grained sediments, and by this, supported the notion of dominance of wind-induced resuspension over other resuspension processes. However, for all wind directions used, there were regions of disparity, with erosional bottoms occurring in areas predicted to be accumulation zones and vice versa. Furthermore, none of the five cases could account for some observed patterns in the spatial distribution of surficial sediment characteristics. In particular, a belt of proximal-distal differentiation in sediment character exists in the northern part of Lake Victoria that cannot be attributed to the action of wind-induced waves. Surficial sediment characteristics further showed no systematic change with increasing water depth or effective fetch. These anomalies, especially the signs of strong directional forcing, are taken to suggest that in Lake Victoria other sediment resuspension processes operate side-by-side with wind-wave resuspension.

The corollary of the above findings is that large lakes, in certain circumstances, behave more like oceans than small lakes. Furthermore, it is clear that predicting sediment resuspension and transport in large lakes from only the action of wind-induced surface waves may leave out other processes that may not alter substantially the position of the mud deposition boundary but still lead to significant redistribution of fine-grained sediments and, hence, to nutrients and pollutants associated with this class of particles. Since generalisations from small lakes may give misleading results when applied to large lakes, it may be necessary that empirical data on large lakes scattered in the body of literature be synthesized and presented in a coherent predictive framework to guide future studies on large lakes.

Regarding sediment focusing, the results clearly show that a focused pattern is not a fixed, invariable outcome of wind-induced resuspension. The basin-wide pattern of sediment distribution appears to be a function mostly of the interaction between wind direction, magnitude and duration on the one hand, and basin morphometry on the other. A focused pattern results when the wind vector is variable in direction during the event that resuspends, transports and redistributes sediment. But where the wind direction is fixed or less variable, sediments are distributed along an upwind-downwind axis. Thus instead of fine sediments accumulating at the lake centre, they accumulate at the (least exposed) upwind end of the lake.

The annual wind cycle over Lake Victoria and other tropical lakes, with unidirectional wind prevailing for several months is quite different from the wind regime over temperate lakes. Over the Laurentian Great Lakes, for example, changes in wind direction accompany the passage of successive high pressure ridges and low pressure troughs that move from west to east with a periodicity of 5-10 days. Severe storms lasting several days are experienced during which the wind vector may rotate through a complete circle (Hamblin and Elder, 1973; Hamblin, 1987;). Such a regime easily produces a focused pattern whereas for regions experiencing unidirectional winds, a directional distribution pattern would seem more like the norm.

The results of the wind analysis are consistent with historical observations on the wind regime in the studied part of the lake. Flohn and Fraedrich (1966) who made one of the earliest studies on atmospheric circulations over Lake Victoria reported that rainfall over the lake is controlled by convergence associated with the nocturnal land-breeze component of the diel pattern of land/lake breezes. They noted that the land breeze regularly produces giant cumulonimbus clusters that, due to prevailing easterly winds, are located during the mature phase over the western part of Lake Victoria, roughly the stretch from Murchison Bay to Bukoba. In this stretch, most rain occurs at night, often in association with strong thunderstorms. This pattern has apparently remained unchanged as indicated by Figure 5.4.

Contrary to the presently held view that southeasterlies have the greatest influence on sediment resuspension and mixing in Lake Victoria (Graham, 1929; Fish, 1957; Talling, 1966), the results of the study suggest that dominant winds come from southerly as well as northerly directions. In Figure 5.3 it was showed that the directional distribution of instantaneous and sustained strong winds was strongly biased towards northerly and southerly directions. Additionally, in Figure 5.15 predictions were presented of the spatial patterns of sedimentological zones due to the two wind directions and it was argued that by combining them, much of the observed pattern in sediment distribution could be explained. Thus the wind regime over Lake Victoria, in my view, is one characterized by co-dominance of the two components of the easterly air currents from the Indian Ocean.

Sediment transport pathways and distribution mechanisms in Lake Victoria
The second objective of this paper was to identify sediment transport pathways and processes in Lake Victoria. The processes could apply to other lakes with similar combinations of morphometric features and meteorological forcing factors. Below, an outline of processes that taken together might explain the observed anomalies and peculiarities in surficial sediment distribution Lake Victoria are offered. The anomalies briefly are: (a) the extreme skews in grain size distribution at most sites; (b) the belt of directional gradation in sediment character; (c) the southward curvature near the lake centre in the spatial distribution of sediment characteristics; and (d) the lack of correlation between surficial sediment properties and fetch and

depth. In the view of this study, a few hydrodynamic processes, the more obvious ones coming to surface in Figure 5.11, link these anomalies.

The factors and processes (some factual, others presumed) taken into consideration are the following:

1. Internal currents occasionally occur and combine with surface waves to intensify bed scour and sediment transport. Wind-induced surface wave action undoubtedly is a major process regulating the resuspension and transport of sediment in the lake. However, it is the belief of this study that at certain times of year, wave action combines with wind-induced internal currents to produce much greater effects on sediment distribution than would be expected from the action of waves alone. The combination, which is common in oceans and possible in large shallow lakes perturbed by strong winds, usually leads to much greater erosion and transport of sediments than is possible when waves alone or currents alone act on the lakebed (Thorn and Parsons, 1980; Lou *et al.*, 2000).

2. Island clusters form a formidable barrier to the movement of wind and lake currents; the largest island chain, the Sese archipelago, probably deflects southerly wind and water currents in a northeastward direction. The Sese archipelago comprises of 66 islands (see Figure 5.1 for island locations) that collectively impeded air-mass movements above the lake surface and water movements below the lake surface. Wave theory predictions showed that southerly strong winds cause sediment resuspension along the entire western coastline and along a SW-NE belt below the Sese and Kome Islands (Figure 5.6d).

3. Nearshore areas on the western part of the lake experience all-year mixing and resuspension. The western coast is downwind of all prevailing wind directions save for the infrequent westerlies. It also is known not to stratify at all (LVEMP, 2002). During the windiest periods, which from Figure 5.4 are March-May and September-November, intensive peripheral wave attack (Hilton *et al.*, 1986) probably occurs extensively over the coast, resuspending and transporting sediment parallel to the shoreline.

4. In the rest of the lake, sediment resuspension and transport probably mainly occurs between June and September. Due to marked offshore stratification (Talling, 1966; Hecky, 1993, Hecky *et al.*, 1994), wind-wave mediated sediment resuspension and transport is unlikely to occur in the rest of the lake in the first part of the year (January-April), notwithstanding the high frequency of strong winds. The main impact of winds between January and April (mostly northerlies) is probably in enhancing evaporative cooling and thermocline-deepening, making it possible for strong southerlies in late May to early June to cause complete overturn. It is plausible that most wind-wave driven sediment distribution takes place in the subsequent period of isothermy between June and September. The minor peak in the frequency of strong winds occurs a few months after annual overturn, between September and November, and is characterized by a relatively high frequency of strong northerly and southerly winds.

5. High-energy hydrodynamic events occasionally occur in the lake. Sediments in the accumulation zone are very poorly sorted and have extremely negatively skewed distributions. This is not unusual as no further sorting is expected to take place once particles cross the mud deposition boundary. However, the presence of heavy coarse fractions (responsible for the extreme negative skews in grain size distribution) in deep offshore accumulation zones is unusual. The thresholds of energy needed to transport medium-to coarse-grained particles are not met by normal wind-driven wave resuspension (Reading and Levell, 1996). Normally, therefore, some degree of segregation should occur that should result in only the lightest particles reaching and crossing the mud deposition boundary.

It may be said from the above factors that sediment resuspension and transport mechanisms other than surface wave resuspension seem to be in operation that deliver sediment of a broad range of particle sizes (fine sand to clay) to a wide area of the lake floor. In the first distribution mechanism offered, an attempt is made to elucidate processes by which a broad range of particle sizes could be widely dispersed in the basin.

Sediment Distribution Mechanism 1
Wind-driven epilimnetic mixing and lake-wide deposition of a homogenous suspension

It is proposed that episodically, there is an influx of sediments of a broad range of particle sizes to the epilimnion that is efficiently mixed under the action of strong prevailing winds, and the resulting uniform suspension is deposition all over the lake in ensuing periods of calm. The sediments dispersed in this manner could come directly from inflowing rivers or from wind-wave resuspension of previously deposited material around basin margins. This process most probably occurs around March-May because of a unique combination of environmental factors at this time of year namely, the occurrence of: (a) the annual peak in offshore stratification producing a well-defined epilimnion; (b) the annual peak in the frequency and magnitude of strong winds that could cause sustained epilimnetic mixing; and (c) the annual peak in rainfall in the catchment that in turn produces annual peaks in river flows and sediment exports. Talling and Lemoalle (1998) argue that the absence of coriolis force at the equator allows greater effectiveness of wind-induced vertical mixing and further cross-isobath penetration of fluvial flows.

There is a second, though less likely, distribution mechanism by which wide dispersal of a broad range of particle sizes could be achieved. This is the process of intermittent complete mixing (Hilton *et al.*, 1986): bottom material is scoured from all areas of the lake and completely mixed by strong sustained winds during periods of isothermy; the uniform suspension then deposits all over the lake floor when calm conditions return. This is less likely to happen on account of the great size of the lake and the long duration of extremely stormy conditions needed to achieve complete lake-wide mixing. In the mechanism suggested above, only the epilimnion, separated from the hypolimnion by a well-developed pycnocline, is mixed.

The above distribution mechanism cannot account for the anomalous SW-NE belt of proximal-distal gradation in sediment character. For this a second mechanism is proposed.

Sediment Distribution Mechanism 2
Current-driven alongshore and cross-isobath transportation of sediments

It is suggested that occasionally, strong southerlies induce alongshore currents on the western coastline that intensify lakebed reworking and transport sediments, both in suspension and as bedload, in northward directions. As the current travels northwards, it eventually meets the barrier of the Sese archipelagos, which splits the alongshore current into two components. The first component continues travelling northwards parallel to the shoreline. A second, and probably larger, component is headed off in the northeast direction. As this deflected current travels further into the lake, its force gradually ebbs. Sediments deposit differentially during transport from the waning of the current, leaving behind a distinct trail of gradational character demarcating the current travel path. The southward curvature in the distribution of sediment characteristics that was observed (Figure 5.9) may be an indication of vorticity in the wind field resulting from wind stress curl caused by the barrier islands (Schwab and Beletsky, 2003).

The critical feature in the second mechanism is the combination of waves and currents in causing sediment redistribution. The river, again, would probably be the primary source of transported sediment, with other material coming from erosion of the lakebed in shallow areas along the western coast. In Figure 5.14, evidence of the Kagera bringing sediments into the lake as bedload was shown. This material is probably first deposited in the nearshore areas before being remobilised and moved further offshore by the combined action of surface waves and currents. The regions of extreme positive skew then represent areas where the power of eroding waves and currents has dissipated to a point where they can no longer keep the heavy fractions in suspension, which then deposit rapidly on the lake floor. According to Folk and Wand (1957), extreme positive skewness in a site indicates that part of the sediment in the site was sorted elsewhere in a high-energy environment before being transported, with its size characteristics essentially unchanged, and redeposited in the site.

The two mechanisms proposed above could be merged if it is assumed that the cross-isobath current is the main conduit for sediments to the epilimnion. It may however be that they operate separately: each at a different time of year.

Evidence for mechanisms

There is considerable geological evidence in support of the above factors and suggested mechanisms, aspects of which have been alluded to by geomorphologists studying the northwestern margins of Lake Victoria (Doornkamp and Temple, 1966; Bishop and Trendall, 1967; Temple, 1969; Temple and Doornkamp, 1970). Temple and Doornkamp (1970) reconstructed the pre-lacustrine topography of the basin and showed it to be the major control upon the subsequent pattern of erosion and sedimentation. They hypothesized that sediments dumped at the ancient delta of the Kagera were subsequently constructed into barriers and spits and progressively transported northwards by wind and waves built up across the open southern part of the lake. The same processes have apparently continued into modern times. Measurements of currents in Lake Victoria that could be used to verify the proposed water flows are unfortunately few and spatially limited.

A second source of evidence comes from multiple core and seismic reflection studies of the lake bottom and its sediments. There are recurring downcore discontinuities in the profile of paleolimnological proxy indicators suggesting episodic strong erosion events that affect large parts of the lake (Johnson et al., 2000; Talbot and Laerdal, 2000; Verschuren et al., 2002). High-resolution seismic reflection profiles show that deposited material in the central-to-eastern parts of the basin is twice as thick as those on the western and southwestern parts, suggesting continuous erosion of material deposited on the western margin (Johnson et al., 1996; Scholz et al., 1998).

The anomalies around site 46 and 62
The anomaly around site 46 may be explained from a consideration of its morphometric features. The site is sandwiched between two small islands - Vumba and Nainaivi - that are part of the Lolui-Dagusi island chain (Figure 5.1). Around the islands, the lakebed is greatly raised giving rise to shallow depths and rocky bottoms. Site 46 has a depth of 10 m and a grain size distribution dominated by coarse particles in sharp contrast to grain size distribution at the nearby site 57 (Figure 5.12). During periods of lake-wide circulation, the loss in volume by the raised bed and large proportion of hard surfaces around the islands could produce high flows sufficient to carry away all but the heaviest particles in previously deposited material in this area.

For the anomalous character of the sediments at site 62 no explanation can be given. This area of the lake is 65-80 m deep but is always extremely turbulent. Current profiles measured at the site with an ADCP (Teledyne RD Instruments, San Diego) between 2000 and 2002 (LVEMP, 2002) showed multidirectional currents over the water column with velocities of $0.15 - 0.25$ m s^{-1}. Near-bed current velocities were often above 0.40 m s^{-1}. The hydrodynamic processes producing such localized high turbulence and a coarse sandy bottom in deep offshore water are still not well understood.

Observations and implications for Lake Victoria
The study findings underscore the urgent need to improve environmental management within the catchment of Lake Victoria. The lake is already severely stressed yet inputs of nutrient-rich sediments from the catchment are continuing, and heavy metal contamination of lake sediments appears to be on the rise (Onyari and Wandiga, 1989; Makundi, 2001; Campbell et al. 2003; Kishe and Machiwa, 2003). The observed pattern of surficial sediment distribution, and above explanations of how they might have arisen, suggest a considerable potential for nutrient recycling and lake-wide dispersal of substances attached to fine-grained sediments. Thus, the lake-wide deterioration in water quality and ecosystem conditions is likely to get worse unless judicious measures are taken to stem the rising trend in pollution.

These results could prove a useful aid in the interpretation of multiple core data from Lake Victoria. During the International Decade for the East African Lakes (IDEAL) (Verschuren et al., 1998) several cores were taken from Lake Victoria two of which (V96-8 and V96-1) lie in the SW-NE belt of gradational sediment character while others lie in the central-to-eastern accumulation zones. The character of sediments in these two zones is markedly different as shown above. Cores from these sites should therefore be able to provide different information, with cores from the SW-NE belt likely to be more sensitive to changes in the wind energy and climate change than the other cores.

The findings of the study appear to be in support of a widely held but yet unproven view in East Africa that the Nile has its headwaters in the highlands of Rwanda and Burundi via the Kagera River and simply passes through Lake Victoria en route the Mediterranean Sea. The data does not confirm the occurrence of a distinct fluvial flow crossing Lake Victoria from the mouth of River Kagera to the outflow of the Nile, a distance of well over 250 km. Still, it would seem from the proximal-distal distribution of sediments that at certain times of year internal currents, whipped up by storms in the lake, help the Kagera to project its load of sediments far into the lake. It is not clear from the limited data how long this flow persists. Probably it is intense but short-lived. This is one area where further research is necessary.

Another area of research is in refinement of the sediment distribution patterns and transport mechanisms proposed here. At the start of the study, a paucity of data made it difficult to justify a larger number of samples than collected. With hindsight, it is recognized that a denser network of sampling sites should have been used. It is recommended therefore that a more detailed sediment study be conducted that should cover the entire lake. The study should be conducted alongside a study of meteorological conditions and currents in the lake.

Acknowledgements

The financial support of the Lake Victoria Environmental Management Project (LVEMP) and logistical support of the Directorate of Water Development (DWD), Uganda is acknowledged with thanks.

References

APHA (American Public Health Association) (1992) In *Standard methods for the examination of water and wastewater.* American Public Health Association, pp. 2-57 to 2-65.

ASTM (American Society for Testing and Materials) (1986) Standard method for particle-size analysis of soils: Method D 422. In *Annual book of ASTM standards.* Vol. 04 Section 4: Construction: 08: Soil and rock; building stones Philadelphia: American Society for Testing and Materials, pp. 116-126.

Bengtsson, L., Hellström, T., and Rakoczi, L. (1990) Redistribution of sediments in three Swedish lakes. *Hydrobiologia* 192: 167-181.

Bishop, W.W. and Trendall, A.F. (1967) Erosion-surfaces, tectonics and volcanic activity in Uganda. *Quarterly Journal of the Geological Society of London* 132: 238-252.

Blais, J.M. and Kalff, J. (1995) The influence of lake morphometry on sediment focusing. *Limnology and Oceanography* 40(3): 582-588.

Campbell, L.M., Hecky, R.E., Nyaundi, J., Mugidde, R., and Dixon, D.G. (2003) Distribution and foodweb transfer of Mercury in Napoleon and Winum Gulfs, Lake Victoria, East Africa. *Journal of Great Lakes Research* 29(2): 267-282.

Carper, G.L. and Bachmann, R.W. (1984) Wind resuspension of sediments in a prairie lake. *Canadian Journal of Fisheries and Aquatic Sciences* 41(12): 1763-1767.

CERC (Coastal Engineering Research Centre) (1984) *Shore protection manual*. Fort Belvoir, Virginia: U.S. Army Corps of Engineers.

Cheng, P., Gao, S., and Bokuniewicz, H. (2004) Net sediment transport patterns over the Bohai Strait based on grain size trend analysis. *Estuarine, Coastal and Shelf Science* 60: 203-212.

Crul, R.C.M. (1995) *Limnology and hydrology of Lake Victoria*. Paris: UNESCO.

Damiani, V. and Thomas, R.L. (1974) The surficial sediments of the Big Bay section of the Bay of Quinte, Lake Ontario. *Canadian Journal of Earth Sciences* 11: 1562-1576.

Denny, P. (1972) The significance of the pycnocline in tropical lakes. *African Journal of Tropical Hydrobiology and Fisheries* 2(2): 85-89.

Doornkamp, J.C. and Temple, P.H. (1966) Surface, drainage and tectonic instability in part of southern Uganda. *The Geographical Journal* 132: 238-252.

Eadie, B.J. and Robbins, J.A. (1987) The role of particulate matter in the movement of contaminants in the Great Lakes. In *Sources and fates of aquatic pollutants*. Hites, R. and Eisenreich, S. (eds). Vol. 216: Washington D. C.: American Chemical Society, pp. 319-364 .

Eadie, B.J., Schwab, D.J., Assel, R.A., Hewley, N., Lansing, M.B., Miller, C.S., Morehead, N.R., Robbins, J.A., and van Hoff, P.L. (1996) Development of recurrent coastal plume in Lake Michigan observed for first time. *Eos, Transactions* 77(35): 337-338.

Evans, R.D. (1994) Empirical evidence of the importance of sediment resuspension in lakes. *Hydrobiologia* 284(1): 5-12.

Evans, R.D. and Rigler, F.H. (1980) Measurement of whole lake sediment accumulation and phosphorous retention using lead-210 dating. *Canadian Journal of Fisheries and Aquatic Sciences* 37(5): 817-822.

Evans, R.D. and Rigler, F.H. (1983) A test of lead-210 dating for the measurement of whole lake soft sediment accumulation. *Canadian Journal of Fisheries and Aquatic Sciences* 40(4): 506-515.

Fish, G.R. (1957) *A seiche movement and its effects on the hydrology of Lake Victoria*. Fisheries Publications, 10 Colon. Off. London. 68 pp.

Flohn, H. and Fraedrich, K. (1966) Tegesperiodische zirkulation und niedersclagsverteilig am Victoria-See (Ostafrica) (The daily periodic circulation and distribution of rainfall over Lake Victoria). *Meteorologische Rundschau* 19: 157-165.

Folk, R.L. and Ward, W.C. (1957) Brazos river bar: a study in the significance of grain size parameters. *Journal of Sedimentary Petrology* 27(1): 3-26.

Gardner, W.D., Richardson, M.J., Hinga, K.R., and Biscaye, P.E. (1983) Resuspension measured with sediment traps in a high-energy environment. *Earth and Planetary Science Letters* 66: 262-278.

Graham, M. (1929) *The Victoria Nyanza and its Fisheries: A report on the fishing surveys of Lake Victoria*. 1929. London, Crown - Agents for the Colonies.

Griffiths, J.F. (1972) Eastern Africa. In *World survey of climatology*. Landsberg, H.E. (ed). Vol. 10: Amsterdam: Elsevier, pp. 313-347.

Guy, H.P. (1994) Laboratory theory and methods for sediment analysis. In *Techniques of water resources investigations of the United States Geological Survey (USGS)*. Book 5: Washington D. C.: USGS, pp. 1-59.

Håkanson, L. (1977) The influence of wind, fetch and water depth on the distribution of sediments in Lake Värnen, Sweden. *Canadian Journal of Earth Sciences* 14: 397-412.

Håkanson, L. and Jansson, M. (1983) *Principles of lake sedimentology*. Berlin: Springer-Verlag.

Hamblin, P.F. (1987) Meteorological forcing and water level fluctuations on Lake Erie. *Journal of Great Lakes Research* 13(4): 435-453.

Hamblin, P.F. and Elder, F.C. A preliminary investigation of the wind stress field over Lake Ontario. Proceedings of the 16th conference on Great Lakes research. 723-734. 1973. International Association of Great Lakes Research.

Harper, D.M. (1992) *Eutrophication of freshwaters: principles, problems and restoration*. New York: Chapman & Hall.

Hawley, N. and Lesht, B.M. (1992) Sediment resuspension in Lake St. Clair. *Limnology and Oceanography* 37(8): 1720-1737.

Hecky, R.E., Bugenyi, F.W.B., Ochumba, P.O.B., Talling, J.F., Mugidde, R., Gophen, M., and Kaufman, L. (1994) Deoxygenation of the deep water of Lake Victoria, East Africa. *Limnology and Oceanography* 39(6): 1476-1481.

Hecky, R.E. (1993) The eutrophication of Lake Victoria. *Verh. Internat. Verein. Limnol.* 25: 39-48.

Hilton, J. (1985) A conceptual framework for predicting the occurrence of sediment focusing and sediment redistribution in small lakes. *Limnology and Oceanography* 30: 1131-1143.

Hilton, J., Lishman, J.P., and Allen, P.V. (1986) The dominant processes of sediment distribution and focusing in a small, eutrophic, monomictic lake. *Limnology and Oceanography* 31(1): 125-133.

Jepsen, R., Roberts, J., and Lick, W. (1997) Effects of bulk density on sediment erosion rates. *Water, Air and Soil Pollution* 99: 21-31.

Johnson, T.C., Scholz, C.A., Talbort, M.R., Kelts, K., Ricketts, R.D., Ngobi, G., Beuning, K.R.M., Ssemmanda, I., and McGill, J.W. (1996) Late Pleistocene desiccation of Lake Victoria and rapid evolution of cichlid fishes. *Science* 273: 1091-1093.

Johnson, T.C. (1984) Sedimentation in large lakes. *Annual Review of Earth and Planetary Sciences* 12: 179-204.

Johnson, T.C., Kelts, K., and Odada Eric (2000) The Holocene history of Lake Victoria. *Ambio* 29(1): 2-11.

Kishe, M.A. and Machiwa, J.F. (2003) Distribution of heavy metals in sediments of Mwanza Gulf of Lake Victoria, Tanzania. *Environment International* 28(7): 619-625.

Leuttich, R.A., Harleman, Jr.D.R.F., and Somyódy, L. (1990) Dynamic behaviour of suspended sediment concentrations in a shallow lake perturbed by wind events. *Limnology and Oceanography* 35(5): 1050-1067.

Lick, W., Lick, J., and Ziegler, C.K. (1994) The resuspension and transport of fine-grained sediments in Lake Erie. *Journal of Great Lakes Research* 20(4): 599-612.

Lou, J., Schwab, D.J., Beletsky, D., and Hawley, N. (2000) A model of sediment resuspension and transport dynamics in southern Lake Michigan. *Journal of Geophysical Research* 105(No. C3): 6591-6610.

LVEMP (Lake Victoria Environmental Management Project). The Lake Victoria Integrated Water Quality and Limnology Study. 2002. Dar-es-Salaam, Tanzania, LVEMP Regional Secretariat.

Makundi, I.N. (2001) A study of heavy metal pollution in Lake Victoria sediments by energy dispersive X-ray fluorescence. *Journal of Environmental Science and Health* A36(6): 909-921.

McLaren, P. (1981) An interpretation of trends in grain size measures. *Journal of Sedimentary Petrology* 51: 611-624.

Newell, B.S. (1960) The hydrology of Lake Victoria. *Hydrobiologia* 15: 363-383.

Nicholson, S.E. (1996) A review of climate dynamics and climate variability in Eastern Africa. In *The limnology, climatology and paleoclimatology of the East African lakes.* Johnson, T.C. and Odada, E. (eds). Amsterdam: Gordon and Breach, pp. 25-56.

Onyari, J.M. and Wandiga, S.O. (1989) Distribution of Cr, Pb, Cd, Zn, Fe and Mn in Lake Victoria sediments, East Africa. *Bulletin of Environmental Contamination and Toxicology* 42: 807-813.

Pond, S. and Pickard, G.L. (1983) *Introductory dynamic oceanography.* Oxford: Pergamon Press.

Reading, H.G. and Levell, B.K. (1996) Controls on the sedimentary record. In *Sedimentary environments: processes, facies and stratigraphy.* Reading, H.G. (ed). Oxford: Blackwell Science, pp. 5-25.

Rowan, D.J., Kalff, J., and Rasmussen, J.B. (1992) Estimating the mud deposition boundary depth in lakes from wave theory. *Canadian Journal of Fisheries and Aquatic Sciences* 49: 2490-2497.

Scholz, C.A., Johnson, T.C., Cattaneo, P., Malinga, H., and Shana, S. (1998) Initial results of 1995 IDEAL seismic reflection survey of Lake Victoria, Uganda and Tanzania. In *Environmental Change and Response in East African Lakes.* Lehman, J.T. (ed). Dordrecht: Kluwer, pp. 47-58.

Schwab, D.J. and Beletsky, D. (2003) Relative effects of wind stress curl, topography, and stratification on large-scale circulation in Lake Michigan. *Journal of Geophysical Research* 108(C2, 3044): doi 10.1029/2001JC001066.

Sheng, Y.P. and Lick, W. (1979) The transport and resuspension of sediments in a shallow lake. *Journal of Geophysical Research* 84: 1809-1826.

Shepard, F.P. (1954) Nomenclature based on sand-silt-clay ratios. *Journal of Sedimentary Petrology* 23(3): 151-158.

Sly, P.G. (1978) Sedimentary processes in lakes. In *Lakes: chemistry, geology physics.* Lerman, A. (ed). New York: Springer Verlag, pp. 65-89.

Talbot, M.R. and Allen, P.A. (1996) Lakes. In *Sedimentary environments: processes, facies and stratigraphy.* Reading, H.G. (ed). Oxford: Blackwell Science, pp. 83-124.

Talbot, M.R. and Lærdal, T. (2000) The Late Pleistocene-Holocene paleolimnology of Lake Victoria, East Africa, based on elemental and isotopic analyses of sedimentary organic matter. *Journal of Paleolimnology* 23: 141-164.

Talling, J.F. (1966) The annual cycle of stratification and phytoplankton growth in Lake Victoria (East Africa). *Internationale Revue der gesamten Hydrobiologie* 51(4): 545-621.

Talling, J.F. and Lemoalle, J. (1998) *Ecological dynamics of tropical inland waters.* Cambridge: Cambridge University Press.

Temple, P.H. Raised strandlines and shorelines evolution in the area of Lake Nabugabo, Masaka District, Uganda. Wright, H. E. Proc. INQUA 16. 1701, 119-129. 1969. National Academy of Sciences.

Temple, P.H. and Doornkamp, J.C. (1970) Influences controlling lacustrine overlap along the north-western margins of Lake Victoria. *Zeitschrift für Geomorphologie* 14(3): 301-317.

Thomas, R.L. (1969) A note on the relationship of grain size, clay content, quartz and organic carbon in some Lake Erie and Ontario sediments. *Journal of Sedimentary Petrology* 39: 803-808.

Thomas, R.L., Jaquet, J.M., Kemp, A.L.W., and Lewis, C.F.M. (1976) Surficial sediments of Lake Erie. *Journal of the Fisheries Research Board of Canada* 33: 385-403.

Thomas, R.L., Kemp, A.L.W., and Lewis, C.F.M. (1972) Distribution, composition and characteristics of the surficial sediments of Lake Ontario. *Journal of Sedimentary Petrology* 42(1): 66-84.

Thomas, R.L., Kemp, A.L.W., and Lewis, C.F.M. (1973) The surficial sediments of Lake Huron. *Canadian Journal of Earth Sciences* 10: 226-265.

Thorn, M.F.C. and Parsons, J.G. Erosion of cohesive sediments in estuaries. Proceedings of the third international symposium on dredging technology. 123-132. 1980. Cranfield, England, BHRA Fluid Eng.

Verschuren, D., Edgington, D.N., Kling, H.J., and Johnson, T.C. (1998) Silica depletion in Lake Victoria: Sedimentary signals at offshore stations. *Journal of Great Lakes Research* 24(1): 118-130.

Verschuren, D., Johnson, T.C., Kling, H.J., Edgington, D.N., Leavit, P.R., Brown, E.T., Talbot, M.R., and Hecky, R.E. (2002) History and timing of human impact on Lake Victoria, East Africa. *Proceedings of the Royal Society of London* 269: 289-294.

Visher, G.S. (1969) Grain size distributions and depositional processes. *Journal of Sedimentary Petrology* 39(3): 1074-1106.

Weyhenmeyer, G.A., Håkanson, L., and Meili, M. (1997) A validated model for daily variations in the flux, origin, and distribution of settling particles within lakes. *Limnology and Oceanography* 42(7): 1517-1529.

Yin, X. and Nicholson, S.E. (1998) The water balance of Lake Victoria. *Hydrological Sciences Journal* 43(5): 789-811.

On the causal factors and mechanisms for ecosystem change in Lake Victoria

Submission to a scientific journal based on this chapter:

Nicholas Azza, Peter Herman, Sabine Schmidt, Johan van de Koppel, Jack J. Middelburg and Patrick Denny. On the causal factors and mechanisms for ecosystem change in Lake Victoria, East Africa, *Global change Biology,* submitted on October 11, 2006.

On the causal factors and mechanisms for ecosystem change in Lake Victoria

Abstract

Striking ecosystem changes have occurred in Lake Victoria that, though not completely understood, are attributed to eutrophication and climate change. To improve understanding of these changes, carbon and nitrogen elemental and isotopic analyses were performed on three short cores taken from Lake Victoria. Sediment age was determined by radiometric (^{210}Pb, ^{137}Cs) dating. The results show that many aspects of the current perceptions on the causal factors and timing of events are correct. A marked increase in the deposition of TOC and TN with concomitant decrease in C/N ratio and δ^{13}C values occurred starting from ca. 1886-1920 that signified an increase in lake primary productivity in response to increasing nutrient loading. There were three areas of discrepancy with current perceptions. First, it was found that the organic matter in the lake was not predominantly of algal origin but comprised of mixtures of land and swamp plants, and phytoplankton. Second, there was no lake-wide synchrony in ecosystem changes, which is presumed to indicate spatial heterogeneity in morphometric, hydrodynamic and water quality conditions. Third, N-fixing cyanobacteria are not dominant over non-N-fixing phytoplankton in far offshore waters. Considering that signal excursions started earlier than the commencement of large-scale anthropogenic disturbance of the watershed, it is argued that eutrophication only amplified a shift in lake conditions that was already underway from climate change. Comparison of ecosystem trends in Lakes Victoria and Tanganyika shows that universal generalizations on the impacts of climate change on aquatic ecosystems may be misleading as contrasting outcomes in lacustrine primary productivity may result from the interaction between impacts of climate change, eutrophication and hydrodynamic regime.

Keywords: Eutrophication, stable isotopes, sediment chronology, organic matter, Lake Victoria.

Introduction

Lake Victoria is the world's second largest freshwater lake by area (Beeton, 1984) and is home to a diverse and unique assemblage of aquatic biota (Coulter et al., 1986). It moderates regional climate, maintains the basic flow of the White Nile and provides drinking water, cheap animal protein, a means of transport and economic livelihood to over 30 million people in East Africa (Crul, 1998). In recent decades tremendous ecological changes have been observed in the lake. Relative to conditions in the early 1960s, concentrations of inorganic nutrients in the water column have risen with the exception of dissolved silica, which has virtually vanished from the lake. Transparency has decreased, algal biomass and photosynthesis have increased, and heterocystous cyanobacteria are believed to have replaced diatoms as the dominant planktonic algae group (Hecky, 1993; Mugidde, 1993, Verschuren et al., 1998; Kling et al., 2001). Coinciding with these changes was an explosion in the population of the piscivorous Nile Perch and the extinction of several hundred species or endemic cichlid fishes (Ogutu-Ohwayo, 1990; Witte et al., 1992). The lake is also appreciably warmer, stratifies more strongly and has an expanded area experiencing prolonged hypolimnetic anoxia (Hecky, 1993; Hecky et al., 1994; Bugenyi and Magumba, 1996).

Theories on the causal factors of the recent ecosystem changes converge on eutrophication and climate change as the primary drivers of the change process (Hecky, 1993; Lehman, 1998; Johnson et al., 2000; Verschuren et al., 2002). The principal elements of these theories are first, that there was rapid human population growth, increased deforestation and slash-and-burn agriculture in the lake's catchment from around 1930, which led to an increase in the export of nutrients to the lake (Hecky, 1993; Verschuren et al., 2002). Second, that phytoplankton (predominantly diatom) productivity increased sharply in response to the rise in nutrient inflows (Lehman and Brandstrator, 1993; Verschuren *et al.*, 2002). Third and last, that the diatom population collapsed in the mid-1980s and was replaced by heterocystous cyanobacteria as the dominant algae group (Hecky, 1993; Mugidde, 1993; Verschuren *et al.*, 2002). The diatom-cyanobacteria switch is attributed to three factors, namely (a) climate change producing weaker winds and greater water column stability leading to increased sinking and burial of diatoms (Hecky and Kling, 1987; Talling, 1987; Lehman, 1998); (b) exhaustion of the lake's dissolved silicon reservoir by increased sinking and burial of silicified algae, and reduced Si recharge from bottom sediments due to greater water column stability (Hecky, 1993, Verschuren *et a.l.*, 1998); and (c) severe nitrogen limitation of algae following increased N-loss through metalimnetic denitrification during extended periods of anoxia (Hecky, 1993). Lake Victoria is generally considered to be N-limited (Talling, 1966; Hecky, 1993; Lehman and Brandstrator, 1993; Mugidde, 1993; Hecky *et al.*, 1996) although parts of the lake exhibit signs of phosphorus limitation, and instances of lake-wide P limitation appear periodically (Mugidde *et al.*, 2003).

Testing and refinement of the above theories is essential but has been limited due to a paucity of limnological and water quality monitoring data. A sound understanding of what exactly happened, and when, how and why it happened could help improve our knowledge of the way in which large tropical lakes respond to nutrient loading and, ultimately, improve our ability to manage them. Lake Victoria has a relatively shallow, flattish and nearly spherical basin (North to South, 400 km; East to West, 210 km; mean depth, 40 m; maximum depth, 84 km) while the other African great lakes, most of them rift valley lakes, have deep, steep-sided basins with more or less oval shapes. New insights on how the interaction between morphometry and hydrodynamic regime influence the response of large lakes to natural and man-made perturbations may be gained by comparing and contrasting ecosystem trends in Lake Victoria against trends in the other African great lakes.

The lack of historical records documenting the ecological changes requires that other methods be used to investigate the transformation. This study uses elemental and isotopic proxies in bulk sedimentary organic matter to reconstruct paleoenvironmental changes in the lake, and identify the processes that drove or participated in the changes. The study examines the way in which different parts of the lake responded to alterations in external forcings, and tests aspects of the above views on the causal factors and mechanisms for ecosystem change in Lake Victoria.

Materials and Methods

Sediment coring

Short cores (45-50 cm long) of diameter 6.7 cm were retrieved in duplicate in June 2004 from three sites along a linear transect starting from the lake margin to the lake centre (Figure 6.1). The first core site (5 m deep) named Northern Nearshore site, is located in a sheltered embayment, Busi Bay, on the northern coast of Lake Victoria. The second site (45 m deep) named Islands site is located north of the sheltering Sese archipelagos. The third site (68 m deep) named Lake Centre is located at the centre of the lake. The cores, which were recovered with a Tech Ops (TOC) gravity corer, were extruded in the field and sampled at 1 cm intervals in the top 0-10 cm section, and 2 cm intervals in the lower 10-40 cm section. Samples were frozen on site, freeze-dried upon return from the field and shipped for elemental and stable isotopic analysis to the Centre for Estuarine and Marine Ecology (CEMO) of the Netherlands Institute of Ecology. Aliquotes of the dried samples were shipped for radiometric (^{210}Pb, ^{137}Cs) dating to the paleolimnology laboratory (UMR EPOC) of the National Centre for Scientific Research (CNRS), University of Bordeaux 1, France.

Elemental and stable isotopic analyses

The content of total organic carbon (TOC wt %) and total nitrogen (TN wt %) were determined using a Carlo-Erba elemental 1500 CN analyser (Carbo-Erba Instruments, Milan) following an *in situ* procedure for inorganic carbon removal involving treatment with HCl (Nieuwenhuize *et al.*, 1994). Isotopic composition of carbon and nitrogen was determined using a Fisons elemental analyzer (Fisons Instruments Inc., Beverley) linked by a continuous flow interface to a Finnigan Delta S mass spectrometer. Results are reported in the conventional delta notation relative to Vienna PDB and atmospheric N_2. Analytical precision determined through replicate measurements was $\pm 0.1\%$ for $\delta^{13}C$ and $\pm 0.13\%$ for $\delta^{15}N$.

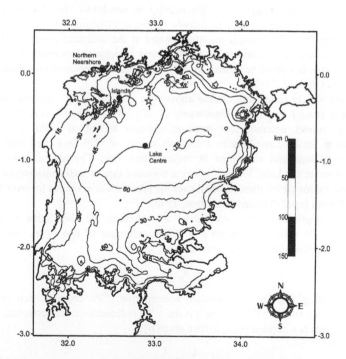

Figure 6.1. Map of Lake Victoria showing bathymetry, core sites (black circles) and algae sampling sites (unfilled stars). Isobaths were obtained by digitising and gridding British Admiralty Charts No. 3252 and 3665 of the Victoria Nyanza. Map coordinates are in decimal degrees.

Sediment geochronology

A combination of ^{210}Pb and ^{137}Cs dating techniques was used. These together provide the most robust approach for dating modern lake sediments (Graham *et al.*, (2004). Under favourable conditions, ^{210}Pb can provide a detailed chronology to *ca.* 120-140 yr, and may be independently validated and accurately calibrated with ^{137}Cs for sediments deposited after 1950.

Total ^{210}Pb activity in sediment comprises of two components termed "supported" and "unsupported". Radioactive decay of ^{226}Ra in the sediment matrix generates the supported ^{210}Pb component via the short-lived daughter nuclide ^{222}Rn while unsupported or excess ^{210}Pb comes from deposition of particles with sorbed ^{210}Pb originating from decay of atmospheric ^{226}Ra (Appleby and Oldfield, 1983).

Different models, depending on environmental forcing factors and the downcore pattern of sediment accumulation, are used to date sediments with ^{210}Pb. In most lakes, a constant flux-constant sedimentation rate (cf:cs) model (Robbins and Edgington, 1975) is used. In these lakes, the flux of sediments from the drainage basin is constant, producing a correspondingly constant rate of sediment accumulation, and an exponential decline in unsupported ^{210}Pb concentrations with increasing sediment depth.

In Lake Victoria and a number of other lakes reported in the literature, the downcore profile of unsupported ^{210}Pb activity is non-linear. Factors cited for deviation from non-linear behaviour include temporal variability in sedimentation rates, exchange of ^{210}Pb through interstitial water at the sediment-water interface, mixing of near-surface sediments by physical and biological processes, and post depositional distribution of sediments by wave-induced resuspension and slumping (Appleby and Oldfield, 1983).

The most widely used model in the above type of lakes is the constant rate of ^{210}Pb supply (c.r.s.) model (Krishnaswamy *et al.*, 1971; Appleby and Oldfield, 1978). This model assumes that there is constant fallout of ^{210}Pb from the atmosphere to the lake resulting in a constant supply of ^{210}Pb to buried sediment irrespective of temporal variations in accumulation rate. The above conditions appear reasonable for Lake Victoria where a constant rate of sedimentation cannot be assumed and for which direct rainfall over the lake surface accounts for over 80% of the total water input (Yin and Nicholson, 1998).

If the assumptions of the c.r.s. model are satisfied, the cumulative residual unsupported ^{210}Pb activity, A_d (dpm g^{-1}), below sediments of age t (yr) will vary according to the formula (Appleby and Oldfield, 1983):

$$A_d = A_o \cdot e^{-\lambda t} \qquad (6.1)$$

where A_o (dpm g^{-1}) is the total residual unsupported ^{210}Pb concentration in the sediment column, and λ (= 0.03114 yr^{-1}) is the ^{210}Pb radioactive decay constant. The age of sediments at a depth d (mm) is then given by:

$$t = \frac{1}{\lambda} \cdot \ln \frac{A_o}{A_d} \qquad (6.2)$$

Least squares regression can then be used to determine a mean linear sedimentation rate from a scatter diagram of sediment age (*t*) versus depth (*d*).

The ^{137}Cs dating technique, which complements ^{210}Pb dating, is applied through identification of marker horizons in the downcore profile of ^{137}Cs activity. The first appearance of ^{137}Cs indicates where 1953 (onset of atmospheric nuclear bomb testing) occurs in the profile provided there is minimal physical disturbance of deposited material. In rain-dominated systems such as Lake Victoria there are additional peaks corresponding to 1963 (the maximum ^{137}Cs fallout due to nuclear weapons testing) and 1986 (Chernobyl nuclear accident). Thus ^{137}Cs dating provides independent absolute time markers for validating the ^{210}Pb chronology that are not affected by bioturbation or other sediment mixing processes (Appleby, 2001; Graham et al., 2004).

The specific activity of radionuclides (^{210}Pb, [226]Ra, ^{137}Cs) was measured by individually counting homogenised samples in a low background, high efficiency well-type germanium gamma detector (Canberra Eurisys, France) (Appleby et al., 1986; Schmidt et al., 2006). The γ detector was calibrated with IAEA reference materials (RGU-1 and RGTh-1). Counting uncertainties associated with sample measurements were typically less than 5% except for low activities of [226]Ra and ^{137}Cs, which had uncertainties up to 8% and 12% respectively. In the absence of atmospheric ^{210}Pb fallout, sedimentary ^{210}Pb and [226]Ra would be in radioactive equilibrium. By subtracting the specific activity of [226]Ra from the ^{210}Pb total specific activity, the unsupported component of ^{210}Pb concentration was obtained and used to determine sediment age using the c.r.s. model as described above.

The age of sediments older than ca. 134 yr (i.e. six half-lives of ^{210}Pb) was derived by downwards extrapolation of the mean sediment accumulation rate obtained in the upper parts of cores, assuming mean sedimentation rates in the lower sections to be comparable to those in the more recent upper sections. Sediment accumulation rates derived from ^{210}Pb dating were generally comparable with rates obtained from ^{137}Cs chronology.

Determining the origin of sedimentary organic matter
Two complementary approaches were used to determine the origin of sedimentary organic matter. In the first, origin was deduced by comparing the signature of the organic matter with respect to molar C_{org}/N_{total} ratio and $\delta^{13}C$ and $\delta^{15}N$ values, to the signature of known groups of primary producers (Hollander and McKenzie, 1991; Schelske and Hodell, 1991; Meyers and Ishiwatari, 1995; Talbot, 2001). Table 6.1 shows the ranges of known groups of primary producers with respect to the above indicators. To improve the predictions of this method, offshore water was sampled (at the time of coring), filtered through GF/F filter (Whatman International, Kent) and trapped material, mainly comprising inorganic suspended sediments and algal biomass, was analysed for the above distinguishing characteristics. The location of algae sampling points is indicated in Figure 6.1. The second approach, described in the section below, uses a numerical fitting procedure and complements interpretations obtained by this method.

Table 6.1. The signature of known plant groups with respect to C/N ratio and $\delta^{13}C$ and $\delta^{15}N$ signatures. Values are from O'Leary (1981), Meyers and Ishiwatari (1995) and Talbot (2001). The midpoints of the ranges have been used in estimating the fractional contributions of the four plants groups indicated in this table.

Parameters	C_3 plants	C_4 plants	Non-N-fixing algae	N-fixing Blue-green algae
$\delta^{13}C$	-30 to -25‰	-14 to -10.5‰	-28 to -25‰	-28 to -25‰
Midpoints	-27.5‰	-12.5‰	-26.5‰	-26.5‰
$\delta^{15}N$	0 to +1‰	0 to +1‰	+6.5 to +8.5‰	0 to +2‰
Midpoints	+0.5‰	+0.5‰	+7.5‰	+1‰
C/N ratio	20-24.5	20-24.5	4-10	4-10
Midpoints	22	22	7	7

Deducing past environmental and limnological changes

In addition to making inferences about the origin of organic matter, the organic geochemical proxies were used to reconstruct the response of the lake to paleoenvironmental changes in the watershed, and interpret the significance of spatio-temporal shifts in the character of organic matter. If it can be assumed that the present condition of the lake is the final outcome of the cause and effect chain described above, then it can be expected, among other things, that (1) the contents of TOC and TN in buried sediments rise markedly after the 1930s; (2) the C/N ratio of buried organic matter rises after the 1930s (3) the $\delta^{13}C$ signal of sedimentary organic matter becomes less negative after the 1930s; (4) the $\delta^{15}N$ signal becomes more positive from the 1930s to mid-1980s and less positive afterwards; or remains unchanged until after the mid-1980s when it becomes less positive; or remains largely unchanged from the 1930s to present times; (5) the elemental and stable isotopic signals of bulk sedimentary organic matter are identical for all parts of the lake, since current suppositions do not indicate differences in the response of the different regions of the lake. The presence or absence, and timing, of excursions in the organic geochemical proxies are used to validate or contest aspects of the above cause and effect chain.

The first four expectations arise from literature interpretations of excursions in organic geochemical proxies, specifically that (a) rising TOC and TN values of bulk sedimentary organic matter usually accompany increases in lake primary productivity (Meyers *et al.*, 1980; Dean 1999; Hodell and Schelske, 1998; Meyers and Teranes, 2001 and Nesje and Dahl, 2001); (b) rising C/N ratios typically denote an increasing supply and burial of land-plant organic matter from growing deforestation and agriculture in the watershed (Tyson, 1995; Kaushal and Binford, 1999; Tenzer *et al.*, 1999); (c) increasing $\delta^{13}C$ values in bulk sedimentary organic matter usually point to periods of high phytoplankton productivity, reflecting reduced discrimination by algae against the heavier ^{13}C isotope as the lighter ^{12}C isotope becomes increasingly scarce (Schelske and Hodell, 1991, 1995; Hodell and Schelske, 1998; Ostrom *et al.*, 1998; Silliman *et al.*, 2001, Meyers, 2003); (d) falling $\delta^{15}N$ values normally indicate increasing proportions of material from N-fixing algae in buried organic matter, and may signify a switch in dominance in the phytoplankton population from non-N-fixing phytoplankton to N-fixing cyanobacteria (Fogel and Cifuentes, 1993; Breener, *et al.*, 1999; Myers and Lallier-Vergès, 1999; Talbot, 2001). The opposite effect to that described i.e. rising $\delta^{15}N$ values, usually implies an increasing proportion of non-N-fixing phytoplankton material in buried organic matter. However, rising $\delta^{15}N$ values may also be caused by increasing inflow of nutrients from terrestrial sources (soil and animal and human wastes) since the NO_3^- in these sources is often isotopically heavier than lacustrine NO_3^- (Kendall, 1998; Teranes and Bernasconi, 2002; O'Reilly *et al.*, 2005), or by increasing denitrification in anoxic bottom waters, leading to production of isotopically heavy N_2 that is subsequently fixed by diazotrophes (Talbot, 2001; Myers, 2003). The latter process, potentially, can mask the emergence of N-fixing cyanobacteria dominance if the isotope content indicator is not combined with other proxies.

There are additional expectations as far as the origin of the organic matter is concerned. It is presently considered (Johnson *et al.*, 2000; Talbot and Lærdal, 2000) that the preserved sedimentary organic matter of Lake Victoria is predominantly of algal origin. The elemental and stable isotopic signatures of the organic matter should accordingly show strong algal character, with major shifts in character only arising from alternations between dominance of non-N-fixing phytoplankton and N-fixing blue-green algae.

One of the aforementioned expectations is the enrichment of the heavy ^{13}C isotope in algal matter with increasing photosynthesis. In a closed system, Raleigh fractionation dictates that ^{13}C fractionation must decrease with increasing CO_2 uptake (Goericke *et al.*, 1994). Where there is an insignificant or unvarying inflow of higher plant organic matter to dilute algal material, it is considered that the $\delta^{13}C$ signature of bulk sedimentary organic matter is controlled chiefly by primary productivity in the water column, and hence, that this parameter can be a good proxy indicator of lacustrine productivity (Schelske and Hodell, 1991; Brenner *et al.*, 1999; Voss *et al.*, 2000; Neumann *et al.*, 2002).

Usually, algae fractionate against ^{13}C during photosynthesis and produce material that is typically depleted in ^{13}C. Under certain circumstances, however, ^{13}C enrichment may occur. First, the lighter ^{12}C isotope (in dissolved CO_2) may be depleted during periods of intense primary productivity and compel cells to rely on the heavier ^{13}C isotope for carbon fixation (Schelske and Hodell, 1991; Neumann *et al.*, 2002). Second, high primary production may strip the water column of all aqueous CO_2 and force plants to switch to HCO_3^- as a carbon source (Hollander and McKenzie, 1991; Keeley and Sandquist, 1992; Bernasconi *et al.*, 1997). In addition, alkaline pH values may cause the establishment of a high HCO_3^- to CO_2 ratio (Hassan *et al.*, 1997). The carbon in HCO_3^- is about 8‰ heavier than carbon in aqueous CO_2; hence the resulting organic matter is ^{13}C-enriched (Espie *et al.*, 1991; Fogel *et al.*, 1992; Meyers, 2003). Lastly, as lakes become more eutrophic, or mix less efficiently, deeper waters experience prolonged anoxia. Under such circumstances, methanogenesis is favoured and produces isotopically light CH_4 and isotopically heavy CO_2 (Stiller and Magaritz, 1974). Thus, in deep stratified large lakes, increasing phytoplankton productivity is normally accompanied by enrichment of the heavy ^{13}C isotope in sedimentary organic matter (Schelske and Hodell, 1991; Voß and Struck, 1997; Lehmann *et al.*, 2004). Exceptions to this general trend have been reported (Bade and Cole, 2006; Bontes *et al.*, 2006;) but are less common.

Fractional composition of organic matter with respect to major plant groups
In a complementary approach to determining the origin of organic matter, the signature of the sedimentary organic matter with respect to molar C/N ratio, δ^{13}C, δ^{15}N and TOC was used to estimate the percentage contributions of inorganic minerals and the four most prominent potential sources or organic matter, that is C_3 plants, C_4 plants, non-nitrogen-fixing algae and N-fixing blue-green algae, in buried sediments. The proportions of organic M_{org} (wt %) and inorganic M_{inorg} (wt %) matter was estimated from the following relationships (Håkanson and Jansson, 1983):

$$M_{org} = 2 \cdot TOC \tag{6.3}$$

$$M_{inorg} = 100 - M_{org} \tag{6.4}$$

The fractional contributions of the above main sources to the sedimentary organic matter was derived by satisfying the following set of equations:

$$CN_{obs} = p_{C_3} \cdot CN_{C_3} + p_{C_4} \cdot CN_{C_4} + p_{Ag} \cdot CN_{Ag} + p_{Bg} \cdot CN_{Bg} \tag{6.5}$$

$$\delta^{13}C_{obs} = p_{C_3} \cdot \delta^{13}C_{C_3} + p_{C_4} \cdot \delta^{13}C_{C_4} + p_{Ag} \cdot \delta^{13}C_{Ag} + p_{Bg} \cdot \delta^{13}C_{Bg} \tag{6.6}$$

$$\delta^{15}N_{obs} = p_{C_3} \cdot \delta^{15}N_{C_3} + p_{C_4} \cdot \delta^{15}N_{C_4} + p_{Ag} \cdot \delta^{15}N_{Ag} + p_{Bg} \cdot \delta^{15}N_{Bg} \tag{6.7}$$

where *CN* represents the C/N ratio of sedimentary organic matter, the subscript *obs* denotes the observed or measured value of a parameter and p_i stands for the fraction of C_3 plants (C_3), C_4 plants (C_4), non-nitrogen-fixing algae (A_g) and N-fixing blue-green algae (B_g) in the organic matter. The values used for the four plant groups for the above distinguishing parameters have been indicated in Table 6.1. Two further constraints were applied as follows:

$$p_i \geq 0 ; \quad \sum p_i = 1 \tag{6.8}$$

This set of conditions results in a linear system of four equations and four unknowns, which were solved numerically using least squares fitting. The obtained values of p_i were multiplied by M_{org} to give the percentage content of the different plant groups.

Reliability of organic geochemical proxies
Since the main conclusions in this study are not based on direct measurements but indirect inferences from organic geochemical proxies, it is essential, before hand, to establish a clear understanding of their reliability, and throw light on uncertainties associated with their use. Microbial mineralisation of organic matter, which starts shortly after formation, and continues after burial of the organic matter, modifies the elemental and isotopic composition of bulk sedimentary organic matter. This

process constitutes the most serious potential source of noise and distortion that could lead to incorrect interpretations.

Several studies have shown that early diagenesis may lead to an increase or decrease in the C/N ratio of suspended and sedimentary organic matter (Freudenthal et al., 2001; Holmer and Olsen, 2002; Killops and Killops, 2005), and may cause either enrichment or depletion in their ^{13}C and ^{15}N contents (Macko and Estep, 1984; Macko et al., 1993; Meyers and Eadie, 1993; Tyson, 1995; Hodell and Schelske, 1998; Ostrom et al., 1998). The principal processes responsible for diagenetic changes are differential loss of labile organic compounds (Macko et al., 1993; Prahl et al., 1997; Freudenthal et al., 2001) and isotopic fractionation during microbial synthesis and mineralisation of organic compounds (Macko and Estep, 1984; Saino, 1992; Macko et al., 1993).

While early diagenetic processes invariably introduce alterations in the chemical nature of buried organic matter, resultant shifts in composition are not considered comparable in magnitude to variations of the primary sedimentary signal. Neither are the changes viewed as sufficiently large to erase completely the differences between major plant groups (Meyers and Ishiwatari, 1995; Hodell and Schelske, 1998; Talbot, 2001; Meyers, 2003). Hence, the above approaches are regarded as sufficiently robust to yield dependable indications on the paleoenvironment of investigated lakes and the composition of bulk sediments with respect to major plant groups.

When using the above applications, it is generally assumed that: (a) the elemental and isotopic compositions of bulk sedimentary organic matter is preserved during transportation in the water column and burial in sediments; (b) the relative proportions of the major constituents are preserved during settling and burial; (c) the contribution of a substance within a mixture, to the stable isotopic signature of the mixture, is a linear function of its fractional composition in the mixture; and (d) the isotopic composition of bulk sedimentary nitrogen reflects mainly the organic source.

Results

Sediment accumulation rates

Profiles of unsupported ^{210}Pb concentrations, were non-linear (Figure 6.2). Good fits ($R^2 > 0.95$) at the three core sites were obtained for linear sedimentation rates using the constant rate of supply model. The calendar dates at the bottom of the cores were estimated as 1820 (Northern Nearshore site), 1725 (Islands site) and 1899 (Lake Centre site). Sediment accumulation rates in the lake are low, with mean rates for the Northern Nearshore, Islands and Lake Centre sites being 2.2 ± 1.0, 1.4 ± 0.6 and 3.8 ± 1.2 mm yr^{-1} respectively. The rate of sedimentation at the Lake Centre site was about twice that at the other sites. In Lake Victoria, as is typical for southern hemisphere systems (Ritchie and McHenry, 1990), instead of a single large peak corresponding to the 1963 ^{137}Cs fallout maximum, there were several peaks of lesser intensity, the largest being around 1964-1966. Ages obtained by ^{210}Pb dating for the ^{137}Cs peaks were in good agreement (Table 6.2), indicating an uncertainty of about 2 years for the last 50 years, and that sediment mixing/bioturbation can be ignored.

Character of sedimentary organic matter

The content of organic carbon in the sediments of Lake Victoria (range 12.3 – 28.1 wt %) is relatively high. Table 6.3 summarizes the character of the sediments at the three core sites with respect to elemental and stable isotopic composition. The vertical profiles of the measured parameters are presented in Figures 6.3-6.5 below.

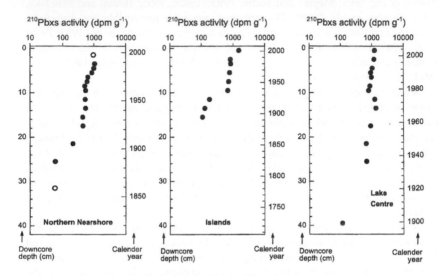

Figure 6.2. Profiles of unsupported ^{210}Pb concentrations showing non-linear patterns. Dates have been derived from a mean sedimentation rate obtained by applying a c.r.s. model to the unsupported ^{210}Pb activity. The points with unfilled circles were disregarded in determination of sediment age.

Table 6.2: Comparison of dates obtained from ^{210}Pb and ^{137}Cs dating methods. The two methods yielded dates that were in reasonably good agreement. The position of the first nuclear weapons test (1953) was not so well defined in the profiles and was not used.

	Event: maximum weapons test (1963)	Event: Chernobyl nuclear accident (1986)
Nearshore Site		
Downcore position of ^{137}Cs peak (cm)	9.5	4.5
Date of peak from ^{210}Pb chronology	1961	1984
Islands Site		
Downcore position of ^{137}Cs peak (cm)	5.5	3.5
Date of peak from ^{210}Pb chronology	1965	1986
Lake Centre Site		
Downcore position of ^{137}Cs peak (cm)	16	6.5
Date of peak from ^{210}Pb chronology	1961	1987

Downcore variation of TOC, TN and C/N ratio

The vertical profiles of TOC and TN are nearly identical, and both look like mirror images of the pattern of C/N ratio. This tight coupling suggests that the three signals are controlled by a common factor, most probably lacustrine primary productivity. This assumption is in accord with current views as it indicates a sharp rise in lake primary productivity to have occurred at *ca.* 1886 in the Nearshore site, *ca.* 1908 at the Islands Sites and *ca.* 1920 at the Lake Centre, and continued to the present day.

Table 6.3. The means and ranges (in parenthesis) of elemental and stable isotopic parameters in bulk sedimentary organic matter and pelagic algae from Lake Victoria. There was very little algal biomass in the water column at the time of sampling. Hence, the content of organic carbon and nitrogen in the filtered residue was much lower than in bottom sediments.

Core (0-40 cm) or sample	TOC (wt %)	TN (wt %)	C/N ratio (mol/mol)	$\delta^{13}C$ (‰)	$\delta^{15}N$ (‰)
Northern Nearshore (n=25)	22.1 [18.1-28.1]	1.9 [1.2-3.3]	14.1 [9.7-18.2]	-20.0 [-21.9 – -17.5]	1.2 [0.6 – 1.8]
Islands (n=25)	15.1 [12.3-18.4]	1.2 [0.9-1.6]	14.5 [12.5-16.6]	-22.1 [-22.7 – -21.6]	2.0 [1.3 – 2.5]
Lake Centre) (n=25)	14.7 [13.0-16.2]	1.1 [0.8-1.5]	16.4 [12.2-18.5]	-22.6 [-24.3 – -22.0]	3.4 [2.7 – 4.0]
Offshore algae: Sample 1	1.27	0.37	7.38	-23.6	3.19
Sample 2	1.55	0.37	7.04	-23.2	4.21

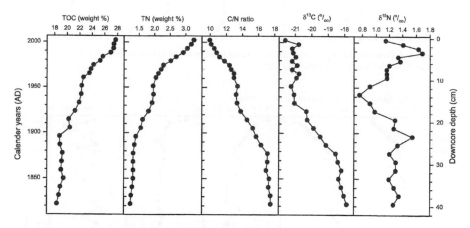

Figure 6.3. Downcore variation of elemental and isotopic composition of bulk sedimentary organic matter at the Northern Nearshore site in Lake Victoria.

The C/N ratio drops in the upcore direction and is negatively correlated with the lake productivity proxy indicator, TOC (Figure 6.6). This pattern in C/N ratio could be interpreted to mean three things, that: (1) as lacustrine primary productivity rises, the proportion of phytoplankton material to higher plant material in sedimentary organic matter increases (organic matter from non-vascular plants contains little or no cellulose and woody tissues, and has relatively large proportions of N-rich proteinaceous materials; consequently, non-vascular plants have low C/N ratios; Meyers and Ishiwatari, 1995; Tenzer *et al.*, 1999; Filley *et al.*, 2001); (2) as nutrient loading from the catchment increases, nitrogen-starvation eases and algal material with higher C/N stoichiometric ratios is produced; or (3) there is a preferential and continuous loss of labile nitrogen in buried organic matter leading to an increasing vascular-plant character in sedimentary organic matter. The TOC – C/N crossplot and results of the origin of organic matter presented below suggest option 1 to be the more likely explanation for the TOC and C/N excursions. The trend in the two proxies rules out the possibility of an increased supply and deposition of higher plant organic matter from the catchment. Such a process would have produced an upcore increase in C/N ratios.

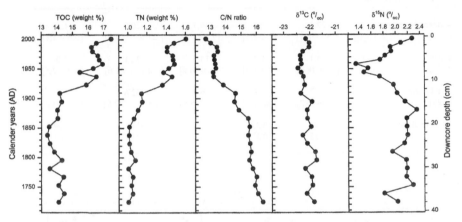

Figure 6.4. Downcore variation of elemental and isotopic composition of bulk sedimentary organic matter at the Islands site in northern Lake Victoria.

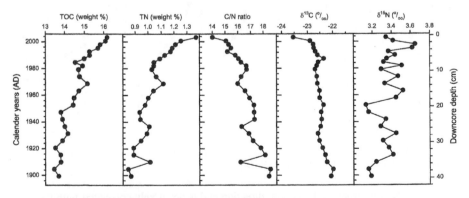

Figure 6.5. Downcore variation of elemental and isotopic composition of bulk sedimentary organic matter at the Lake Centre site in Lake Victoria.

The change in lake primary productivity, inferred from the above proxy indicators, apparently did not proceed at a uniform rate but alternated between phases of rapid and slow change. Different parts of the lake also responded differently to the stimulus of increased nutrient inflows. The nearshore site differed from the two offshore sites, which were similar in many respects. In the nearshore site (Figure 6.3), the rise in TOC and TN with concomitant drop in C/N ratio started a decade earlier than at the offshore sites (Figures 6.4-6.5) and registered the greatest net increases. There appear to be two instances when the rate of change slowed down considerably (1942-1960 and 1988-2004). The offshore sites at which rapid change commenced at *ca.* 1908-1920 were characterized by a single period during which there was either stagnation (at Islands site; 1937-1979) or regression (at Lake Centre, 1968-1984). The changes at the sheltered Islands site lags behind the other sites, and combined with the hiatus in the ^{210}Pbxs profile (Figure 6.2) and low sedimentation rates, perhaps reflect periodic erosion or non-deposition at this site. The most rapid shift in signals occurred in the Nearshore Site from 1965-1988. After what seems like a lag of two decades, the rapid shift manifested in the top 1995-2004 section of the Lake Centre core.

It is possible that some of the trends in TOC, especially in the Lake Centre site, are due to temporal variations in dilution of organic matter with mineral sediments, a reflection of varying sedimentation conditions through time. The inflections in the TOC profiles correspond to sections of constant values or negative inflexions in unsupported ^{210}Pb activity, which is indicative of episodes of higher sediment accumulation.

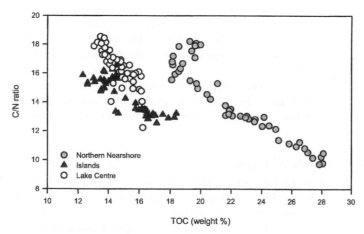

Figure 6.6. Cross plot of TOC and C/N ratio of bulk sedimentary organic matter in cores from Lake Victoria.

Downcore variation of the $\delta^{13}C$ signal

At the three sites, there is a general depletion in the heavy ^{13}C isotope moving from the bottom to the top of the cores. At the Northern Nearshore site (Figure 6.3), the rate of depletion was fairly rapid before 1950, and relatively slower thereafter. In the offshore sites (Figures 6.4 and 6.5), the shift was more gradual, and the rate of decline nearly even through the core.

The excursions in TOC, TN and C/N ratio above indicate an increase in phytoplankton productivity in the three sites commencing at different periods. From theoretical expectations (see above), this process should have been accompanied by enrichment in the heavy ^{13}C isotope. However, the observed trend is depletion in the ^{13}C content. If the organic material in the sediment were to be entirely composed of algal matter as argued by Talbot and Lærdal (2000) and embraced by Johnson *et al.*, (2000), this trend would suggest a tendency for algae to increasingly discriminate against the heavier isotope as algal photosynthesis intensifies. This conflicts with current generalizations mentioned above on $^{13}C/^{12}C$ fractionation during phytoplankton photosynthesis.

If, in contradiction to the above views, it can be assumed that bulk sedimentary organic matter in the lake is not exclusively comprised of algal material but is a mixture of higher plant and phytoplankton material, then the depletion in the ^{13}C signal signifies an increasing proportion of phytoplankton material in a mixture containing C_4 plant material. Bulk sedimentary organic matter in the Northern Nearshore site is richer in the heavier ^{13}C isotope than material in the other sites, most likely reflecting a higher content of C_4 plant material due to proximity of the site to papyrus-dominated fringe swamps. In addition, the magnitude of change in the $\delta^{13}C$ value, which indicates the degree of stimulation of phytoplankton productivity, was largest (2.5‰) at the Northern Nearshore site. This, again, could have been due to the closeness of the site to land and, hence, land-derived nutrients.

Also discernible in the $\delta^{13}C$ signals, besides the general decline in $\delta^{13}C$ values, are brief episodes of ^{13}C enrichment. These occur from 1950-1965 (Northern Nearshore site), 1970-1985 (Island Site) and 1980-1988 (Lake Centre site), and presumably reflect periods of very high phytoplankton productivity during which acute shortage of the lighter ^{12}C isotope leads to synthesis of algae material that is highly enriched in the ^{13}C isotope.

In all sites, there is enrichment in the $\delta^{13}C$ content in the upper 2 cm of core. The 0.5 – 1.5‰ shift could be pointing to alterations in isotopic composition introduced by microbial reworking of sedimentary organic matter during early diagenesis.

Downcore variation of the $\delta^{15}N$ signal

The $\delta^{15}N$ signal in the three sites underwent a series of rises and falls. At the Northern Nearshore site (Figure 6.3), the first major shift, a rapid depletion in the heavy ^{15}N isotope coinciding with the period of rapid increase in TOC and TN (proxy indicators of lake primary productivity) and rapid decrease in C/N ratios, occurred from 1896-1942. The minimum $\delta^{15}N$ values correspond to a temporary halt in the downward C/N trend. This shift could imply (a) a rise in N-fixing cyanobacteria productivity leading to an increase in the proportion of material from N-fixing phytoplankton in buried organic matter; or (b) an increase in the supply of higher plant material from surrounding lands and fringe swamps leading to an increase in the proportion of material from vascular plants in buried organic matter. The previous sections rule out the possibility of the latter occurrence. Hence, this shift in the $\delta^{15}N$ signal is probably due to increasing productivity by N-fixing cyanobacteria. Increasing productivity by non-N-fixing phytoplankton would have produced enrichment in the $\delta^{15}N$ signal while increasing loading of land and swamp plant materials would have produced an increase in C/N ratio.

The second major shift in the $\delta^{15}N$ value at the Northern Nearshore site, enrichment in the heavy ^{15}N isotope, occurred in two steps from 1942-1959, and 1969-1970. This reversal may be attributed to (a) an increase in the productivity of non-N-fixing phytoplankton; (b) a decrease in reliance on nitrogen fixation by N-fixing cyanobacteria; (c) a rising proportion of nitrogen from soil and human and animal wastes (with isotopically heavy NO_3^-) in the pelagic nitrogen pool; or (d) an increase in hypolimnetic or benthic denitrification producing isotopically heavy N_2. The character of organic matter at the Northern Nearshore site (Table 6.3) shows a strong dominance of blue-green algae, which would suppress the growth of other phytoplankton. Therefore, options (b) to (d) appear the more likely causes for the shift in the $\delta^{15}N$ value. The increased supply of soil-derived nitrogen is perhaps the more important of the above processes and, particularly, could have dominated in the period following the 1960s rise in lake levels and drowning of marginal lands and swamps.

The third major shift at the Northern Nearshore site, depletion in the heavy ^{15}N isotope, occurs in the upper 3.5 cm of core and probably signifies diagenetic modifications in the composition of buried organic matter.

The general trend in the $\delta^{15}N$ profile at the offshore sites is similar to that at the Northern Nearshore site. At the Islands site (Figure 6.4), the profile is characterised by a rapid depletion in the heavy ^{15}N isotope coinciding with a rapid shift in TOC, TN and C/N ratio from 1880-1958, followed by a rapid enrichment in the ^{15}N isotope from 1958-2004. In the Lake Centre site (Figure 6.5), the profile looks somewhat different from that at the other sites due to the multiple rises and falls in the proxy indicator. However, the overall trend in the $\delta^{15}N$ signal is not fundamentally different from that at the other sites. There is first a small depletion in the heavy ^{15}N isotope from 1915-1952 followed by a general enrichment from 1952-2004. These trends, as suggested above, point to periods of increased N-fixing cyanobacteria productivity succeeded by periods of increased supply of isotopically heavy NO_3^- and N_2 from soils, sewage or hypolimnetic denitrification. Current presumptions thus seem to be flawed in as far as they consider that in the first half of the twentieth century the response of all areas of the lake to increased nutrient inflow was increased diatom productivity.

At the offshore sites, there is no indication of diagenetic modification of the composition of buried organic matter as clearly evidenced at the Northern Nearshore site. The offshore sites are also generally less enriched in the heavier ^{15}N isotope indicating smaller proportions of N-fixing cyanobacteria material in buried sediments at these sites than at the Northern Nearshore site.

Origin of the organic matter and nearshore-offshore and bottoms-up trends in origin
To verify and further elucidate the cause for the shifts in the elemental and stable isotopic signals, the signature of organic matter as summarized in Table 6.3 has been graphically compared against known plant groups in Figure 6.7. The comparison, in concurrence with earlier deductions, suggests that the organic matter of Lake Victoria has a character between that of phytoplankton and higher plants i.e. the sedimentary organic matter is most likely a mixture of material from algae and terrestrial and wetland macrophytes.

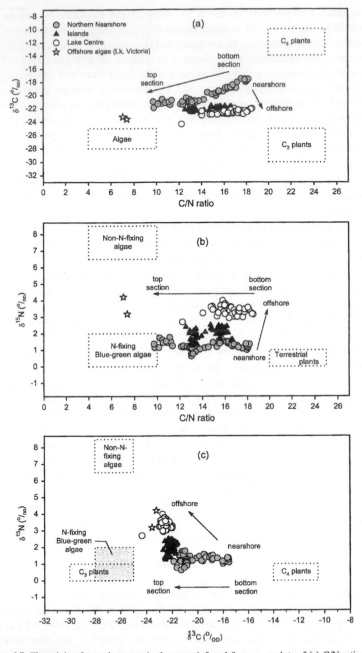

Figure 6.7. The origin of organic matter in the cores inferred from cross plots of (a) C/N ratio and δ¹³C signature; and (b) C/N ratio and δ¹⁵N signature; and (c) δ¹³C versus δ¹⁵N signatures. The doted rectangles, which use values from Meyers and Ishiwatari (1995), indicate the typical range of values for C/N ratio and stable isotopic composition for known plant groups. The arrows show the general bottom-up and nearshore-offshore trends in organic matter character.

Arrows showing bottoms-up and nearshore-offshore trends in character have been included in Figure 6.7. Moving upwards through the cores, there is a progressive shift towards a more algal character that, in conformity with current views, suggests a progressive increase in lacustrine primary productivity in response to presumed increases in nutrient inflows. Moving from nearshore to offshore, there is switch in dominance from C_4 plant material to non-N-fixing algae material.

The above interpretations are closely corroborated and strengthened by the outcome of the second approach (Figure 6.8), which used a numerical fitting procedure to estimate the fractional composition of the sediments with respect to the major plant groups. The main inferences to be drawn from Figure 6.8 are that: (1) the character of organic matter in the lake is variable in time and space; (2) at the Northern Nearshore site, prior to *ca.* 1886, bulk sedimentary organic matter was dominated by input from vascular C_4 plants (probably material washed out from papyrus-dominated fringe swamps); after *ca.* 1886, the contribution of vascular plants decreased while that of N-fixing blue-green algae increased; (3) at the offshore sites, prior to *ca.* 1900, organic matter in buried sediments had roughly equal proportions of vascular and non-vascular material (the vascular material had a strong C_3 plant character); after *ca.* 1900, the contribution of vascular plants remained almost unchanged while the contribution of phytoplankton material in the mixture increased; the change at the Islands site was due to a greater contribution of N-fixing blue-green algae while at the Lake Centre, it was due to a greater contribution by non-nitrogen-fixing algae; (4) moving from nearshore to offshore, there is a decrease in the contribution of N-fixing blue-green algae, and increase in the contribution of other phytoplankton, in buried organic matter.

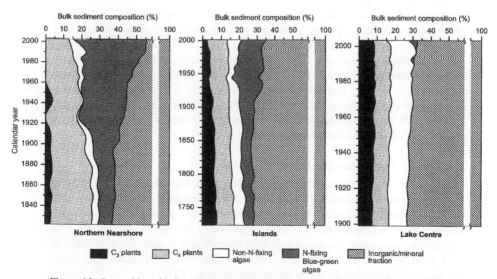

Figure 6.8. Composition of bulk sediments in the cores deduced by modeling techniques from the stable isotopic and elemental signatures of sedimentary organic matter, and from the total organic carbon content.

Discussion

Summary of findings

This study compared the paleolimnological record preserved in bottom sediments with expectations from current perceptions regarding the recent limnological changes in Lake Victoria. The results verify several aspects of these conceptions, contradict others, and bring to light new findings on the processes surrounding the ecological transformation of the lake. The organic geochemical record presented here is supportive of the view that (a) there was an increased inflow of nutrients from the catchment, especially from soils and animal and human wastes; (b) there was increased phytoplankton productivity over much of the previous century in response to the increased inflow of nutrients; (c) there were periods when the rate of increase in phytoplankton productivity was moderate and when it was rapid; and (d) denitrification, probably from prolonged hypolimnetic anoxia, increased after the 1950s.

While the results uphold the view of eutrophication as a driver of ecosystem change in Lake Victoria, disparity between the sediment record and current views was noted in three areas. First, the data suggests that the sedimentary organic matter in the lake is a mixture of material from terrestrial plants, fringe swamp macrophytes and lake phytoplankton rather than being exclusively of algal origin. Secondly, there was no lake-wide synchrony in the commencement and conclusion of elemental and stable isotopic signal shifts, and no homogeneity in limnological changes. Thirdly, at the Lake Centre site, which is taken to represent far offshore areas, the amount of N-fixing cyanobacteria material in the most recent sediments is not greater than that of non-N-fixing phytoplankton.

A number of new findings emerged in the study, namely that (i) stimulation in phytoplankton productivity by presumed nutrient inflows was greatest at the nearshore area and decreased towards the lake centre; (ii) the response of nearshore and intermediate-offshore sites to eutrophication was mainly through increased N-fixing cyanobacteria productivity while that of far offshore sites was through increased non-N-fixing phytoplankton productivity; and (iii) in the nearshore and intermediate-offshore sites there was no change in dominance of phytoplankton groups: N-fixing cyanobacteria dominated the phytoplankton population in both types of sites throughout the studied period.

The dominant phytoplankton group in offshore waters

The signature of all non-N-fixing phytoplankton with respect to $\delta^{13}C$, $\delta^{15}N$ and C/N ratio is nearly identical. Consequently, the geochemical proxies used in this study are unable to distinguish between the different groups of non-N-fixing phytoplankton and cannot estimate the contribution of blue-green algae to the total phytoplankton biomass. The proxy indicators are only able to show the proportion of total algal biomass that is comprised by N-fixing blue-green algae, yet many members of this division do not fix nitrogen (Fogg et al., 1973).

The above limitation may be responsible for the apparent disparity between these results and previous studies indicating collapse of diatoms and emergence of blue green dominance in the offshore algal population (Hecky, 1993; Mugidde, 1993, Kling et al., 2001; Verschuren et al., 2002). The far offshore record, consistent with the previous studies, indicates a first appearance around 1990, and rising proportion thereafter, of N-fixing algae material in sedimented organic matter. However, according to the record, the amount of N-fixing algae material in the sediments is not proportionally larger than the combined contribution of non-N-fixing algae. How can this be reconciled with the above indications of cyanobacteria takeover in the 1990s?

Cyanobacteria dominance of the primary producer population of Lake Victoria is unquestionable. However, it would seem that the cyanobacteria assemblage in the lake has a large proportion of non-N-fixing species, and the combined biomass of the non-N-fixing algae (which not only includes non-N-fixing cyanobacteria but also diatoms, green algae, dinoflagellates, euglenophytes, etc) is roughly equivalent to, or greater than, that of N-fixing cyanobacteria. It is possible that earlier works underestimated the combined biomass of non-N-fixing species. It is also plausible that for part of the time N-fixing species in the lake rely on NH_4^+ and NO_3^- instead of N_2 as a source of nitrogen.

Scientists have recently challenged the notion that cyanobacteria dominate in eutrophic lakes because of their ability to fix nitrogen where the N:P supply ratio is low, and nitrogen is a limiting nutrient (Ferber et al., 2004). The workers showed for a bloom-prone lake that cyanobacteria dominance could occur even when the predominant source of N is dissolved inorganic nitrogen. They found symptoms of both light and nitrogen limitation in the lake, and noted low heterocyst densities confirming low reliance on N-fixation. Ferber et al. (2004) have suggested that cyanobacteria may become dominant by forming surface scums that shade out other algae, and monopolizing benthic sources of ammonia. Moreover, Scholten et al. (2005) and Marino et al. (2006) have shown that grazing by zooplankton may suppress development of N-fixing cyanobacteria in conditions of low N:P ratios. Thus, it is possible that the reliance on N-fixation in Lake Victoria is not so heavy despite the lake being nitrogen limited.

Comparison of Lakes Victoria and Tanganyika

Lake Tanganyika, a tropical great lake also located in East Africa, presents a suitable case against which to compare and contrast findings of this study. The comparison allows insightful inferences to be made about the interaction between climate change, eutrophication, hydrodynamic regime and lacustrine primary productivity. Both lakes are under the influence of the same regional environmental and climatic controls (Spigel and Coulter, 1996). The qualitative impacts of global climatic change in the two lakes have also been the same. The main impacts, which have arisen as a consequence of reduced windiness, warming trend in water temperatures and stronger thermal stratification, have been reduced vertical mixing, shrinking epilimnion and increasing area experiencing prolonged or permanent hypolimnetic anoxia (Hecky, 1993; Hecky et al., 1994; O'Reilly et al., 2002). Furthermore, both lakes have experienced increased sediment and nutrient inflows from anthropogenic disturbance of the catchment over the last century (Hecky, 1993; Cohen et al., 2005). However, the trends in lacustrine primary productivity have been in opposite directions. In Lake Victoria, there was a rise in primary productivity with increasing nutrient inflows whereas in Lake Tanganyika, there was a decline in primary productivity with increasing nutrient inflows (Verburg et al.,

2003; O'Reilly *et al.* 2003). What could be the reason for this disparity and what inferences can be drawn from these two cases about the impacts of climate change and eutrophication?

The differences in lacustrine primary productivity most probably arise from differences in the frequency and spatial extent of nutrient upwelling and entrainment in surface waters. Internal loading of deep-water nutrients is an important nutrient source for the pelagic zone in the two lakes (Hecky *et al.*, 1991; 1996). The differences in internal loading are in turn probably related to differences in morphometry. Lake Victoria has a shallow, flattish and roughly spherical basin that makes it possible for strong winds blowing from any direction to induce basin-wide mixing. Lake Tanganyika, on the other hand, has a deep, narrow and elongated basin that can only be mixed by winds blowing along its longitudinal axis. Its large depth (mean depth, 570 m; maximum depth, 1470 m) further makes it impossible to achieve complete mixing by mechanical means. Thus in Lake Victoria nutrients supplied from the catchment are episodically returned to the water column where they stimulate greater primary production while in Lake Tanganyika they mostly end up in the hypolimnion and in bottom sediments. In a recent study in which the wind regime and spatial pattern of surficial sediments was examined (Azza *et al.*, in prep.) strong evidence was found for the episodic occurrence of severe storms in Lake Victoria and for lake-wide recharge of nutrients from bottom sediments during the storms. In Lake Tanganyika, entrainment of nutrient-rich bottom waters in surface waters takes place through thermocline tilting (O'Reilly *et al.*, 2002; 2003) but clearly is of a much smaller scale than in Lake Victoria.

The implication from the above discussion is that predicting the impacts of climate change on lacustrine ecosystems can be difficult because modifying factors can lead to contrasting outcomes in different aquatic systems. This in itself is not a new finding but is nicely illustrated by the comparison of lakes Tanganyika and Victoria.

Climate change and ecosystem transformation
One question that remains unanswered in looking at the vertical profiles of paleolimnological proxies is whether the increase in nutrient inflows to Lake Victoria since *ca.* 1920 (1) introduced a shifting trend in lake conditions from a previous position of relative stability; (2) speeded up change in the direction of a previous natural trend; (3) reversed the direction of change of a previous natural trend; or (4) cancelled out the effect of a previous natural trend. It would appear from the vertical elemental and stable isotopic signals, especially the trends at the Lake Centre core, that option (2) is the more likely answer i.e. cultural eutrophication did not initiate a new trend but speeded up an existing shift in lake conditions. The change in all five parameters in the Lake Centre core (Figure 6.5) clearly commenced much earlier than the period of study (*ca.* 1880-2004). The same can be said for C/N ratio and the δ^{13}C signal in the Northern Nearshore core, and for C/N ratio in the Islands core (Figures 6.3 and 6.4). Human populations in the lake catchment before then were most probably too low to be responsible for the observed signal shifts. That the recent trends are a continuation of earlier trends becomes more apparent when paleolimnological data for longer (> 1000 yr) time spans are examined (see for example Talbot and Lædal, 2000; and Johnson *et al.*,

2000). Therefore, it may be imperative to look beyond the catchment, towards more regional or global-scale trends in climate conditions for a more satisfactory explanation of the signal excursions.

Timing and progression of changes

A general pattern of change was detectable in the downcore profiles of measured parameters but there were differences in the timing and in the magnitude and rates of signal changes in the three sites. In Lake Tanganyika, such differences, reinforced by regional differences in dominant phytoplankton, were attributed to spatial differences in hydrodynamic regime and nutrient recharge (Descy *et al.*, 2005). These same differences in Lake Victoria may therefore be pointing to considerable spatial heterogeneity in morphometric, hydrodynamic and physico-chemical conditions in Lake Victoria and, hence, to different initial states and buffering capacities in the different regions of the lake.

It was observed that, generally, impacts first manifest in the nearshore zone before gradually being transmitted towards the lake centre. The increase in TOC and TN with simultaneous decline in molar C/N ratios and $\delta^{13}C$ content commenced at the Northern Nearshore site about half a century before it appeared in the offshore sites, and the phase of most rapid change appeared in the Lake Centre site two decades after it's commencement in the Northern Nearshore site. It is important to bear in mind that given the limitations in core resolution and uncertainties associated with dating, the timing differences may not be so significant. Nevertheless, the data does seem to portray a trend of change from lake margin towards lake centre.

The above discussion can be further developed to guide the making of choices between nearshore and offshore sites in paleolimnological work. The observed trends suggest that nearshore sites may preserve a better record of the influence of catchment disturbances than far offshore sites. The proximity of nearshore areas to the inflow point of terrigenous material implies that they are more impacted than far offshore sites where impacts are tempered by dilution effects. This view is supported by observations from deltaic studies in Lake Tanganyika (O'Reilly *et al.*, 2005), which have shown that the proportion of allochthonous material in bulk sedimentary organic matter decreases with distance from shore. The proximity of nearshore areas also means there will be a smaller time lag between cause and effect, making it easier to match demographic and land use changes in the catchment with response preserved in bottom sediments. However, the potential for nearshore areas preserving a more detailed record is usually countered by less favourable sedimentation conditions, unless the selected area is sufficiently deep or adequately sheltered. Then again, in such a case, the record will reflect a local history that cannot be extrapolated to the whole basin. Thus, the offshore site may be a better choice for coring where the study objective is reconstruction of the basin-wide history of a lake.

Organic matter preservation in Lake Victoria

The content of organic carbon in the recent sediments of Lake Victoria is unusually high and far greater than the organic carbon content in sediments of similarly sized lakes (Einsele *et al.*, 2001). The organic carbon content was high even in the period prior to the 1880s i.e. before the changes related to cultural eutrophication. This suggests that the high organic content of the sediments is not simply a reflection of the occurrence of eutrophication (though it was clearly enhanced by eutrophication) but to the existence of conditions favourable for formation of organic-rich deposits and underlines a potential for tropical aquatic systems to play an important role in the sequestration of atmospheric carbon.

The main requirements for formation of organic-rich deposits and preservation of the composition of organic matter are: (a) a large supply of organic material of terrestrial and/or lacustrine origin; (b) a relatively small proportion of inorganic mineral material in sedimentary matter; (c) a high rate of linear accumulation; (d) a large proportion of low-energy depositional bottoms; (e) occurrence of anoxia in bottom waters; and (f) a relatively small proportion of materials recycled by detritivores and decomposers (Hodell and Schelske, 1998; Einsele et al., 2001; Killops and Killops, 2005). The high content of organic carbon in the sediment of Lake Victoria may be indicative of the occurrence of one or several of the above requirements, and for relatively intact preservation of sedimentary organic matter in the lake.

A reasonably large export of plant material from surrounding land, one of the requirements for intact preservation of organic matter, is plausible for Lake Victoria. The lake lies within the equatorial climatic belt of Africa that is characterized by high rainfall and warm, humid conditions. Its catchment area is about three times its surface area and has a large cover of tropical forests, woodlands and savannahs. Also present in the catchment and along its shorelines are large tropical wetlands dominated by highly productive plants such as papyrus. Thus, higher plant debris may be continually flowing in from the catchment and supplementing autochthonous production to give the organic-rich character in deposits of Lake Victoria. Indeed on inspection of the debris deposited in the sub-aqueous delta of River Kagera (the largest inflowing river) a profusion of terrestrial and swamp plant fragments and root mat shreds was noted. The interpretation of this study (presented in preceding sections), of the signature of core material with respect to elemental and stable isotopic composition supports such a view. It has been suggested that (1) sedimentary organic matter prior to ca. 1896 was predominantly of terrestrial and swamp origin and that (2) with increasing nutrient inflows to the lake especially after ca. 1920, algal productivity in the water column was stimulated producing in turn an enhanced downward flux of algal detritus and progressive shift towards a strongly algal character in sedimentary organic matter.

The infrequence of high-energy mixing events and dominance of low-energy depositional bottoms could also be contributing factors to the high organic content of the sediments of the lake. Analysis of wind data for 1994-2004 (Azza et al., in prep.) found weak winds, typically 3-4 m s^{-1}, to occur for much of the time over the lake although there are episodic occurrences of strong storms. The same study found about 70% of the lake bottom in the northern half of Lake Victoria to be overlain by fine-grained sediments. Such conditions favour, or are indicative of suitable conditions for, the formation of organic rich deposits.

Spatial variation in rates of sediment accumulation
In agreement with previous investigations of the lake floor (Johnson et al., 1996; Scholz et al., 1998) and recent studies on sediment distribution patterns (Azza et al.,

in prep.), the variability in sediment accumulation rates that was noted in the sites suggests that most sediment is transported and deposited in the deep offshore region at the centre of the lake. High resolution seismic reflective profiles of the lakebed found significant differences in the total thickness of sediment across the basin, with deposition preferentially concentrated in the northeastern third of the lake, and little accumulation in other parts. Studies of surficial sediment characteristics found fine-grained sediments to accumulate mostly in the central to eastern parts of the lake.

Acknowledgements

Fieldwork in Uganda was funded by the Lake Victoria Environmental Management Project (LVEMP). Initiation of inter-institutional collaboration was facilitated by a START (IHDP, IGBP and WCRP program) Visiting Scientist Award to Nicholas Azza in 2002.

References

Appleby, P.G. (2001) Chronostratigraphic techniques in recent sediments. In: *Tracking Environmental Change Using Lake Sediments*, W.M. Last and J.P. Smol (eds). Kluwer Academic Publishers, Dordrecht pp. 171-203.

Appleby, P.G., Nolan, P.J., Gifford, D.W., Godfrey, M.J., Oldfield, F., Anderson, N.J., and Batterbee, R.W. (1986) 210Pb dating by low background gamma counting. *Hydrobiologia*, 143: 21-27.

Appleby, P.G., and Oldfield, F. (1978) The calculation of lead-210 dates assuming a constant rate of supply of unsupported ^{210}Pb to the sediment. *Catena*, 5: 1-8.

Appleby, P.G., and Oldfield, F. (1983) The assessment of ^{210}Pb data from sites with varying sediment accumulation rates. *Hydrobiologia*, 103: 29-35.

Azza, N.G.T., Koppel, v.d.J., and Denny, P. (In prep.) Pattern and mechanisms of sediment distribution in Lake Victoria, East Africa.

Bade, D.L., and Cole, J.J. (2006) Impact of chemically enhanced diffusion on dissolved inorganic carbon stable isotopes in a fertilised lake. *Journal of Geophysical Research*, Vol 111, CO1014, doi:10.1029/204JC002684.

Beeton, A.M. (1984) The world's Great Lakes. *Journal of Great Lakes Research*, 10(2): 106-113.

Bernasconi, S.M., Barbieri, A., and Simona, M. (1997) Carbon and nitrogen isotope variations in sedimenting organic matter in Lake Laguno. *Limnology and Oceanography*, 42: 1755-1765.

Bontes, B.M., Pel, R., Ibelings, B.W., Boschker, H.T.S., Middelburg, J.J., and Van Donk, E. (2006) The effects of biomanipulation of the biogeochemistry, carbon isotopic composition and pelagic food web relations of a shallow lake. *Biogeosciences*, 3: 69-83.

Brenner, M., Whitmore, T.J., Curtis, J.H., Hodell, D.A., and Schelske, C.L. (1999) Stable isotope (δ^{13}C and δ^{15}N) signatures of sedimented organic matter as indicators of historic lake trophic state. *Journal of Paleolimnology*, 22: 205-221.

Bugenyi, F.W.B., and Magumba, K.M. (1996) The present physico-chemical ecology of Lake Victoria, Uganda. In: *The Limnology, Climatology and Paloeclimatology of the East African Lakes*, T.C. Johnson and E.O. Odada (eds). Gordon and Breach, Amsterdam pp. 141-154.

Cohen, A.S., Palacios-Fest, M.R., Msaky, E.S., Alin, S.R., McKee, B., O'Reilly, C.M., Dettman, D.L., Nkotagu, H., and Lezzar, K.E. (2005) Paleolimnological investigations of anthropogenic environmental change in Lake Tanganyika: IX. Summary of paleorecords of environmental change and catchment deforestation at Lake Tanganyika and impacts on the Lake Tanganyika ecosystem. *Journal of Paleolimnology*, 34: 125-145.

Coulter, G.W., Allanson, B.R., Bruton, M.N., Greenwood, P.H., Hart, R.C., Jackson, P.B.N., and Ribbink, A.J. (1986) Unique qualities and special problems of African Great Lakes. *Env. Biol. Fish.*, 17: 161-184.

Crul, R.C.M. (1998) *Management and Conservation of the African Great Lakes*, UNESCO, Paris 107 pp.

Dean, W.E. (1999) The carbon cycle and biogeochemical dynamics in lake sediments. *Journal of Paleolimnology*, 21: 375-393.

Descy, J.-P., Hardy, M.-A., Stenuite, S., Pirlot, S., Leporcq, B., Kimirei, I., Sekadende, B., Mwaitega, S.R., and Sinyenza, O. (2005) Phytoplankton pigments and community composition in Lake Tanganyika. *Freshwater Biology*, 50: 668-684.

Einsele, G., Yan, J., and Hinderer, M. (2001) Atmospheric carbon burial in modern lake basins and its significance for the global carbon budget. *Global and Planetary Change*, 30: 167-195.

Espie, G.S., Miller, G.A., Kandasamy, R.A., and Cavin, D.T. (1991) Active HCO_3^- transport in cyanobacteria. *Canadian Journal of Botany*, 69: 936-944.

Ferber, L.R., Levine, S.N., Lini, A., and Livingston, G.P. (2004) Do cyanobacteria dominate in eutrophic lakes because they fix atmospheric nitrogen? *Freshwater Biology*, 49(6): 690-708.

Filley, T.R., Freeman, K.H., Bianchi, T.S., Baskara, M., Colarusso, L.A., and Hatcher, P.G. (2001) An isotopic biogeochemical assessment of shifts in organic matter input to holocene sediment from Mud Lake, Florida. *Organic Geochemistry*, 32: 1153-1167.

Fogel, M.L., and Cifuentes, L.A. (1993) Isotope fractionation during primary productivity. In: *Organic Geochemistry, Principles and Applications*, M.H. Engel and S.A. Macko (eds). Plenum, New York pp. 73-98.

Fogel, M.L., Cifuentes, L.A., Velinsky, D.J., and Sharp, J.H. (1992) Relationship of carbon availability in estuarine phytoplankton to isotope composition. *Mar. Ecol. Prog. Ser.*, 82: 291-300.

Fogg, G.E., Stewart, W.D.P., Fay, P., and Walsby, A.C. (1973) *The Blue-Green Algae*, Academic Press, London 459 pp.

Freudenthal, T., Wagner, T., Wenzhöfer, F., Zabel, M., and Wefer, G. (2001) Early diagenesis of organic matter from sediments of eastern subtropical Atlantic: evidence from stable nitrogen and carbon isotopes. *Geochimica et Cosmochimica Acta*, 65(11): 1795-1808.

Goericke, R., Montoya, J.P., and Fry, B. (1997) Physiology of isotopic fractionation in algae and cyanobacteria. In: *Stable Isotopes in Ecology and Environmental Science*, K. Lajtha and R.H. Michener (eds). Blackwell Publishers, Oxford pp. 187-221.

Graham, I.J., Ditchburn, R.G., and Barry, B.J. (2004) ^{210}Pb - ^{137}Cs dating of glacial lake sediments. *New Zealand Science Review*, 61(2): 45-47.

Hakanson, L., and Jansson, M. (1983) *Principles of Lake Sedimentology*, Springer-Verlag, Berlin 316 pp.

Hassan, K.M., Swinehart, J.B., and Spalding, R.F. (1997) Evidence of Holocene environmental change from C/N ratios and $\delta^{13}C$ and $\delta^{15}N$ values in Swan Lake sediments, western Sand Hills, Nebraska. *Journal of Paleolimnology*, 18: 121-130.

Hecky, R.E., Bootsma, H.A., Mugidde, R., and Bugenyi, F.W.B. (1996) Phosphorus pumps, nitrogen sinks and silicon drains: Plumbing nutrients in the African great Lakes. In: *The Limnology, Climatology and Paloeclimatology of the East African Lakes*, T.C. Johnson and E.O. Odada (eds). Gordon and Breach Publishers, Amsterdam.

Hecky, R.E., Bugenyi, F.W.B., Ochumba, P.O.B., Talling, J.F., Mugidde, R., Gophen, M., and Kaufman, L. (1994) Deoxygenation of the deep water of Lake Victoria, East Africa. *Limnology and Oceanography*, 39(6): 1476-1481.

Hecky, R.E., and Kling, H.J. (1987) Phytoplankton ecology of the great lakes in the rift valleys of Central Africa. *Arch. Hydrobiol. Beih.*, 25: 197-228.

Hecky, R.E., Spigel, R.H., and Coulter, G.W. (1991) The nutrient regime. In: *Lake Tanganyika and Its Life*, G.W. Coulter (ed). Oxford University Press, Oxford pp. 76-89.

Hecky, R.E. (1993) The eutrophication of Lake Victoria. *Verh. Internat. Verein. Limnol.*, 25: 39-48.

Hodell, D.A., and Schelske, C.L. (1998) Production, sedimentation and isotopic composition of organic matter in Lake Ontario. *Limnology and Oceanography*, 43(2): 200-214.

Hollander, D.J., and McKenzie, J.A. (1991) CO_2 control on carbon-isotope fractionation during aqueous photosynthesis: a paleo-pCO_2 barometer. *Geology*, 19: 929-932.

Holmer, M., and Olsen, A.B. (2002) Role of decomposition of mangrove and seagrass detritus in sediment carbon and nitrogen cycling in a tropical mangrove forest. *Marine Ecology Progress Series*, 230: 87-101.

Johnson, T.C., Scholz, C.A., Talbot, M.R., Kelts, K., Ricketts, R.D., Ngobi, G., Beuning, K.R.M., Ssemmanda, I., and McGill, J.W. (1996) Late Pleistocene desiccation of Lake Victoria and rapid evolution of cichlid fishes. *Science*, 273: 1091-1093.

Johnson, T.C., Kelts, K., and Odada, E.O. (2000) The Holocene history of Lake Victoria. *Ambio*, 29(1): 2-11.

Kaushal, S., and Binford, M.W. (1999) Relationship between C:N ratios of lake sediments, organic matter sources, and historical deforestation of lake Pleasant, Massachusetts, USA. *Journal of Paleolimnology*, 22: 439-442.

Keeley, J.E., and Sandquist, D.R. (1992) Carbon: freshwater plants. *Plant and Cell Environment*, 15: 1021-1035.

Kendall, C. (1998) Tracing nitrogen sources and cycles in catchments. In: *Isotope Tracers in Catchment Hydrology*, Kendall C. and J.J. McDonnell (eds). Elsevier, New York pp. 519-576.

Killops, S.D., and Killops, V.J. (2005) *Introduction to Organic Geochemistry*, Blackwell Publishing, Malden, USA 393 pp.

Kling, H.J., Mugidde, R., and Hecky, R.E. (2001) Recent changes in the phytoplankton community of Lake Victoria in response to eutrophication. In: *The Great Lakes of the World (GLOW): Food-Web, Health and Integrity*, M. Munawar and R.E. Hecky (eds). Backhuys, Leiden, The Netherlands pp. 47-65.

Krishnaswamy, S., Lal, D., Martin, J.M., and Meybeck, M. (1971) Geochronology of lake sediments. *Earth and Planetary Science Letters*, 11: 407-414.

Lehman, J.T., and Brandstrator, D.K. (1993) Effects of nutrients and grazing on phytoplankton of Lake Victoria. *Verh. Internat. Verein. Limnol.*, 25: 850-855.

Lehman, J.T. (1998) Role of climate in the modern condition of Lake Victoria. *Theor. Appl. Climatol.*, 61: 29-37.

Lehmann, M.F., Bernasconi, S.M., McKenzie, J.A., Barbieri, A., Simona, M., and Veronesi, M. (2004) Seasonal variation of the d13C and d15N of particulate and dissolved carbon and nitrogen in Lake Lugano: Constraints to biogeochemical cycling in a eutrophic lake. *Limnology and Oceanography*, 49(2): 415-425.

Macko, S.A., Engel, M.H., and Parker, P.L. (1993) Early diagenesis of organic matter in sediments: assessment of mechanisms and preservation by the use of isotopic molecular approaches. In: *Organic Geochemistry, Principles and Applications*, M.H. Engel and S.A. Macko (eds). Plenum, New York pp. 211-224.

Macko, S.A., and Estep. M.L.F. (1984) Microbial alteration of stable nitrogen and carbon isotopic compositions of organic matter. *Organic Geochemistry*, 6: 787-790.

Marino, R., Chan, F., Howarth, R.W., Pace, M.L., and Likens, G.E. (2006) Ecological constraints on planktonic nitrogen fixation in saline estuaries. I. Nutrient and trophic controls. *Marine Ecology Progress Series*, 309: 25-39.

Meyers, P.A., Bourbonniere, R.A., and Takeuchi, N. (1980) Hydrocarbons and fatty acids in two cores of Lake Huron sediments. *Geochimica et Cosmochimica Acta*, 44: 1215-1221.

Meyers, P.A., and Eadie, B.J. (1993) Sources, degradation and recycling of organic matter associated with sinking particles in Lake Michigan. *Organic Geochemistry*, 20: 47-56.

Meyers, P.A., and Ishiwatari, R. (1995) Organic matter accumulation records in lake sediments. In: *Physics and Chemistry of Lakes*, A. Lerman, D.M. Imboden and J.R. Gat (eds). Springer-Verlag, Berlin, Germany pp. 279-328.

Meyers, P.A., and Lallier-Vergés, E. (1999) Lacustrine sedimentary organic records of Late Quaternary paleoclimates. *Journal of Paleolimnology*, 21(3): 345-372.

Meyers, P.A., and Teranes, J.L. (2001) Sediment organic matter. In: *Tracking Environmental Changes Using Lake Sediments Vol. II: Physical and Geochemical Methods*, W.M. Last and J.P. Smol (eds). Kluwer, Dordrecht.

Meyers, P.A. (2003) Applications of organic geochemistry to paleolimnological reconstructions: a summary of examples from the Laurentian Great Lakes. *Organic Geochemistry*, 34: 261-289.

Mugidde, R., Handzel, L., Hecky, R.E., and Taylor, W.D. (2003) Pelagic nitrogen fixation in Lake Victoria (East Africa). *Journal of Great Lake Research*, 29(2): 76-88.

Mugidde, R. (1993) The increase in phytoplankton primary productivity and biomass in Lake Victoria (Uganda). *Verh. Internat. Verein. Limnol.*, 25: 846-849.

Nesje, A., and Dahl, S.O. (2001) The Greenland 8200 cal. yr BP event detected in loss-on-ignition profiles in Norwegian lacustrine sediment sequences. *Journal of Quaternary Science*, 16(2): 155-166.

Neumann, T., Stögbauer, A., Walpersdorf, E., Stüben, D., and Kunzendorf, H. (2002) Stable isotopes in recent sediments of Lake Arendsee, NE Germany: response to eutrophication and remedial measures. *Palaeogeography, Palaeoclimatology, Palaeoecology*, 178: 75-90.

Nieuwenhuize, J., Maas, Y.E.M., and Middelburg, J.J. (1994) Rapid analysis of organic carbon and nitrogen in particulate materials. *Marine Chemistry*, 44: 217-224.

O'Leary, M. (1981) Carbon isotope fractionation in plants. *Phytochemistry*, 20(4): 553-567.

O'Reilly, C.M., Dettmen D.L., and Cohen, A.S. (2005) Paleolimnological investigations of anthropogenic environmental change in Lake Tanganyika VI: Geochemical indicators. *Journal of Paleolimnology*, 34: 85-91.

O'Reilly, C.M., Hecky, R.E., Cohen, A.S., and Plisnier, P.-D. (2002) Interpreting stable isotopes in food webs: Recognizing the role of time averaging at different trophic levels. *Limnology and Oceanography*, 47(1): 306-309.

O'Reilly, C.M., Alin, S.R., Plisnier, P.-D., Cohen, A.S., and McKee, B.A. (2003) Climate change decreases aquatic ecosystem productivity of Lake Tanganyika, Africa. *Nature*, 424: 766-768.

Ogutu-Ohwayo, R. (1990) The decline of the native fishes of Lake Victoria and Kyoga (East Africa) and the impact of introduced species, especially the Nile perch, *Lates niloticus*, and the Nile tilapia, *Oreochromis niloticus*. *Environmental Biology of Fishes*, 27: 81-86.

Ostrom, P.H., Ostrom, N.E., Henry, J., Eadie, B.J., Meyers, P.A., and Robbins, J.A. (1998) Changes in trophic state of Lake Erie: discordance between molecular and bulk $\delta^{13}C$ sedimentary records. *Chemical Geology*, 152: 163-179.

Prahl, F.G., de Lange, G.J., Scholten, S., and Cowie, G.L. (1997) A case of post-depositional aerobic degradation of terrestrial organic matter in turbidite deposits from the Madeira Abyssal Plain. *Organic Geochemistry*, 27: 141-152.

Ritchie, J.C., and McHenry, J.R. (1990) Applications of radioactive fallout Cesium-137 for measuring soil and sediment accumulation rates and patterns: a review. *Journal of Environmental Quality*, 19: 215-233.

Robbins, J.A., and Edgington, D.N. (1975) Determination of recent sedimentation rates in Lake Michigan using ^{210}Pb and ^{137}Cs. *Geochimica et Cosmochimica Acta*, 39: 285-304.

Saino, T. (1992) ^{15}N and ^{13}C natural abundance in suspended particulate organic matter from a Kuroshio warm-core ring. *Deep-Sea Research*, 39: 347-362.

Schelske, C.L., and Hodell, D.A. (1995) Using carbon isotopes of bulk sedimentary organic matter to reconstruct the history of nutrient loading and eutrophication in Lake Erie. *Limnology and Oceanography*, 40 : 918-929.

Schelske, C.L., and Hodell, D. (1991) Recent changes in productivity and climate of Lake Ontario detected by isotopic analysis of sediments. *Limnology and Oceanography*, 36: 961-975.

Schmidt, S., Jouanneau, J.-M., Weber, O., Lecroart, P., Radakovitch, O., Gilbert, F., and Jezequel, D. (2006) Sediment dynamics of reworking at the sediment-water interface of the Thau Lagoon (South France); From seasonal to century time scales using radiogenic and cosmogenic nuclides. *Estuarine Coastal Shelf Science*: In Press.

Scholten, M.C.T., Foekema, E.M., Van Dokkum, H.P., Kag, K.N.H.B.M., and Jak, R.G. (2005) *Eutrophication Management and Ecotoxicology*, Springer-Verlag, New York 200 pp.

Scholz, C.A., Johnson, T.C., Cattaneo, P., Malinga, H., and Shana, S. (1998) Initial results of 1995 IDEAL seismic reflection survey of Lake Victoria, Uganda and Tanzania. In: *Environmental Change and Response in East African Lakes*, J.T. Lehman (ed). Kluwer, Dordrecht pp. 47-58.

Silliman, J.E., Meyers, P.A., Eadie, B.J., and Klump, J.V. (2001) An hypothesis for the origin of perytene based on its low abundance in sediments of Green Bay, Wisconsin. *Chemical Geology*, 177: 309-322.

Spigel, R.H., and Coulter, G.W. (1996) Comparison of hydrology and physical limnology of the East African Great Lakes: Tanganyika, Malawi, Victoria, Kivu and Turkana (with reference to some North American Great Lakes). In: *The Limnology, Climatology and Paloeclimatology of the East African Lakes*, T.C. Johnson and E.O. Odada (eds). Gordon and Breach Publishers, Amsterdam pp. 103-140.

Stiller, M., and Magaritz, M. (1974) Carbon-13 enriched carbonate in interstitial waters of Lake Kinneret sediments. *Limnology and Oceanography*, 19: 849-853.

Talbot, M.R. (2001) Nitrogen Isotopes in Palaeolimnology. In: *Tracking Environmental Change Using Lake Sediments*, W.M. Last and J.P. Smol (eds). Kluwer Academic Publishers, Dordrecht, The Netherlands pp. 401-439.

Talbot, M.R., and Lærdal, T. (2000) The Late Pleistocene-Holocene paleolimnology of Lake Victoria, East Africa, based on elemental and isotopic analyses of sedimentary organic matter. *Journal of Paleolimnology*, 23: 141-164.

Talling, J.F. (1987) The phytoplankton of Lake Victoria (East Africa). In: *Phycology of Large Lakes of the World*, M. Munawar (ed). *Arch. Hydrobiol. Beih. Ergebn. Limnol.*, pp. 25:229-256.

Talling, J.F. (1966) The annual cycle of stratification and phytoplankton growth in Lake Victoria (East Africa). *Internationale Revue der gesamten Hydrobiologie*, 51(4): 545-621.

Tenzer, G.E., Meyers, P.A., Robbins, J.A., Eadie, B.J., Morehead, N.R., and Lansing, M.B. (1999) Sedimentary organic matter record of environmental changes in the St Marys River ecosystem, Michigan-Ontario border. *Organic Geochemistry*, 30: 133-146.

Teranes, J.L., and Bernasconi, S.M. (2000) The record of nitrate utilisation and productivity limitation provided by $\delta^{15}N$ values in lake organic matter - a study of sediment trap and core sediments from Baldeggersee, Switzerland. *Limnology and Oceanography*, 45: 801-813.

Tyson, R.V. (1995) *Sedimentary Organic Matter: Organic Facies and Palynofacies*, Chapman and Hall, London 615 pp.

Verburg, P., Hecky, R.E., and Kling, H. (2003) Ecological consequences of a century of warming in Lake Tanganyika. *Science*, 301(5632): 505-507.

Verschuren, D., Edgington, D.N., Kling, H.J., and Johnson, T.C. (1998) Silica depletion in Lake Victoria: Sedimentary signals at offshore stations. *Journal of Great Lakes Research*, 24(1): 118-130.

Verschuren, D., Johnson, T.C., Kling, H.J., Edgington, D.N., Leavit, P.R., Brown, E.T., Talbot, M.R., and Hecky, R.E. (2002) History and timing of human impact on Lake Victoria, East Africa. *Proceedings of the Royal Society of London*, 269: 289-294.

Voss, M., Larsen, B., Leivuori, M., and Vallius, H. (2000) Stable isotope signals of eutrophication in Baltic Sea sediments. *Journal of Marine Systems*, 25: 287-298.

Voß, M., and Struck, U. (1997) Stable nitrogen and carbon isotopes as indicator of eutrophication of the Oder River (Baltic Sea). *Marine Chemistry*, 59(1): 35-49.

Witte, F., Goldschmidt, T., Wanink, J., van Oijen, M., Goudswaard, K., Witte-Maas, E., and Bouton, N. (1992) The destruction of an endemic species flock: quantitative data on the decline of the haplochromine cichlids of Lake Victoria. *Environmental Biology of Fisheries*, 34: 1-28.

Yin, X., and Nicholson, S.E. (1998) The water balance of Lake Victoria. *Hydrological Sciences Journal*, 43(5): 789-811.

Shoreline wetlands and lacustrine sediments: A general discussion

Shoreline wetlands and lacustrine sediments:
A general discussion

Introduction

In the previous chapters, the results of five studies were presented and discussed. In this final analysis, key findings that help advance our understanding of the ecological dynamics of lake shoreline systems in general, and Lake Victoria in particular, are discussed. Areas for future research, specifically examining speculations and hypotheses put forward by this research, are proposed.

Reflections on some themes

The distribution of shoreline wetlands: simplistic versus complex approaches
According to ecological theory (Hutchson, 1975; Spence, 1982; Scheffer, 1998; Keddy, 2000) the distribution of shoreline wetlands is regulated by environmental factors such as water depth, light intensity, substratum type, hydrological regime, wind-wave action, and by disturbance and inter-specific competition for space and resources. From the multiplicity of factors, the occurrence of shoreline vegetation is the net result of a complex interplay between divergent and convergent influences. In spite of this, a number of studies, such as Keddy (1985), simplify this process, attributing it to a few key factors. A simplistic approach has the advantage of minimising data collection costs and, provided all critical influences are included, it should suffice for most purposes of predicting and accounting for the spatial pattern of shoreline vegetation. However, simplification may blind the researcher to the contribution of relatively passive influences that, in changed circumstances, may assume great importance. Interaction between driving forces often results in amplification or cancellation of each other's influences. Thus, focussing on what are perceived as critical influences without allowing for possible interaction between factors may lead to greater or lesser importance assigned to variables than should be the case, and to failure to explain certain aspects of the within-lake distribution of aquatic vegetation.

Chapters 2 and 3 of this thesis confirm general theory by demonstrating that wave exposure, investigated through a proxy (maximum effective fetch), is a major regulator of the presence and size of shoreline wetlands. The thesis, in addition, revealed the influence of a couple of obscure underlying factors whose contributions augment the influence of well-known regulating factors. From the study of the influence of wave exposure on shoreline wetland distribution, it became apparent that bay characteristics had an influence on the distribution of shoreline wetlands. I have speculatively suggested that the real underlying influences are the geometry of the bays and wave-topography interactions. Previous studies have shown that the shape of bays can result in wave convergence or divergence, and that nearshore bathymetry and alongshore structures including shoreline wetlands strongly affect the cross-shore propagation of waves (Holman and Bowen, 1979; Mei, 1989; Clouet and Fouque, 1994; Maas *et al.*, 1997; Petrevu and Oltman-Shay, 1998; Nachbin and

Solna, 2003). These studies suggest the occurrence of feedbacks between forcing factors and vegetation establishment, which I have alluded to in this thesis. The beach geometry and within-bay propagation of waves is different before and after the establishment of shoreline wetlands in a lake. As wetlands establish along a beachface, the open-water geometry and cross-shore propagation of waves will change leading to a modified impact of wind and waves.

In agreement with general theory (Hutchinson, 1975; Spence, 1982; Bertness, 1999; Keddy, 2000), and expanding on the list of commonly cited regulating factors, an important role for disturbance in controlling the distribution of the shoreline wetlands of Lake Victoria is implied in Chapter 3. In the littorals of the lake, plants with a dual growth form (i.e. having bottom-rooted and floating-mat growth forms) have a dominant status. The thesis speculatively presents the dual growth form as having a pivotal role in the influence of disturbance. I argue that possession of a dual growth form gives plants tolerance over a greater range of environmental conditions, allowing them to survive large fluctuations in lake level and become pioneer colonizers following major disturbances. This viewpoint needs to be followed up for a more complete understanding of the controls on the occurrence of shoreline plants in aquatic systems.

Alternative stable states in nearshore waters

The potential for multiple stable states in freshwater systems complicates the interpretation of variability in space and time of the presence and size of lacustrine wetlands, and can limit our ability to predict ecosystem responses to natural and man-made environmental change. Most studies on this phenomenon have been conducted in small and isolated lakes (for example see Scheffer, 1990; Blindow *et al.*, 1993; and Lowe *et al.*, 2001) although multiple stable states have been proposed for more open systems (see for example Knowlton, 1992; Van de Koppel *et al.*, 2001). Additionally, previous studies have been conducted mostly in the middle and high latitude regions. This study provides indication that processes within nearshore areas of large deep lakes are similar to the processes within small or shallow lakes. The results presented in Chapters 4 and 6 suggest that the overall response of the nearshore regions of Lake Victoria to nutrient enrichment is similar to that in many smaller lakes. Although little direct evidence is given for the presence of alternative stable states, it is proposed that feedbacks associated with alternative stable states will severely impair the potential for recovery of the shoreline wetlands of Lake Victoria.

In small shallow lakes, euhydrophytes compete with phytoplankton for space and resources, the dominance of one or the other being a self-perpetuating stable state (Scheffer, 1990, 1998; Jense, 1997; Scheffer and Jeppesen, 1998). A high density of algae in water can shade out and ultimately eliminate euhydrophytes. Once the switch to phytoplankton dominance has occurred, there is resistance to re-establishment of euhydrophytes. So strong can the resistance be that reductions in external nutrient loads may fail to cause reversal in states (Jeppesen *et al.*, 1991).

The alternative euhydrophyte-dominated state, which is characteristic of oligotrophic systems, is also self-stabilising. In such systems, euhydrophytes actively suppress the development of phytoplankton dominance through five mechanisms, namely: (1) resisting flow of water with their submerged body parts thereby allowing inflowing streams to drop their load of suspended matter (Dieter, 1990; James and Barko, 1990; Petticrew and Kalff, 1992); (2) binding and stabilizing bottom sediments thus preventing their resuspension (Meyer *et al.*, 1990; Kufel and Ozimek, 1994; Van den Berg *et al.*, 1998); (3) uptaking nutrients from the water column with their submerged organs in this way maintaining low nutrient concentrations in the water (Denny, 1972; Van Donk *et al.*, 1983; Mothes, 1987; Madsen and Cedergreen, 2002); (4) releasing allelopathic substances into the water that suppress the growth of phytoplankton (Wium-Andersen *et al.*, 1982; Wium-Andersen, 1987; Nikai *et al.*, 1999, 2000; Van Donk and Van de Bund, 2002); and (5) sheltering Daphnia and other pelagic cladocerans from fish predation, and allowing them to graze on phytoplankton (Timms and Moss, 1984; Jeppesen *et al.*, 1997; Scheffer, 1999; Norlin *et al.*, 2005). By these mechanisms, euhydrophytes keep turbidity and algal biomass levels down and underwater light levels up. Thus, when not degraded, the euhydrophytes act to sustain their dominance and, with it, the intricate ecosystem dynamics and valuable functions of food, nursery grounds, refuge and sexual isolation to fish and other aquatic fauna.

It is argued in earlier Chapters that prior to eutrophication, shoreline systems in Lake Victoria were euhydrophyte-dominated. A delicate natural balance existed that allowed euhydrophytes to co-exist with emergent floating swamps. It is speculated that with increased nutrient loading from the catchment, coupled with greater flushing of shoreline swamps, greater shading by algae and humic substances occurred leading to reduction in euhydrophyte cover and a switch to a phytoplankton-dominated state with adverse consequences for fish.

Besides eutrophication, man-made water releases resulting in rapid rises and falls in water level may bring about the destruction of shoreline vegetation and deprive fish and other aquatic fauna of breeding, nursery and feeding grounds. With a lowering in water level, the old vegetated shoreline will slowly transform into terrestrial land but at the new shoreline, macrophytes will not establish immediately. Therefore, for Lake Victoria, large falls in water level and shifts in shoreline will be accompanied by an initial boom in Nile perch populations given the ease with which the predator will capture prey in the absence of camouflaging macrophyte structures. However, this will be short-lived as the prey population, unable to recover rapidly because of lack of suitable nursery grounds, will fall sharply causing a serious decline in the Nile perch fishery.

Sediment distribution in great lakes: wind-induced resuspension versus current resuspension mechanisms

The size of a water body has strong implications for its hydrodynamic and sedimentological characteristics. In small lakes, wind-induced resuspension is the only important sediment transport mechanism (Bengtsoon *et al.*, 1990; Leuttich *et al.*, 1990; Rowan *et al.*, 1992, Evans, 1994,) while in oceans, current-mediate transport mechanisms assume an equally important role to wind-induced distribution mechanisms (Nielsen, 1992; Reading and Levell, 1996; Thurman and Trujillo, 2004). How would sediments be transported in a system that is too large to be a small lake but too small to be an ocean? Simple logic leads us to expect that such a system would have a character between that of small lakes and oceans. Indeed, this is the case as the thesis has shown.

In Chapter 5, examination of the spatial pattern of surficial sediment distribution led to the conclusion that wind-induced sediment resuspension is the dominant sediment distribution mechanism in Lake Victoria. However, the study found evidence of strong influence of currents on sediment distribution. On the basis of these findings, I have argued that in certain circumstances, great lakes behave more like small lakes, while in other circumstances they behave more like oceans. I have further suggested that a focused pattern of sediment distribution is not a foregone conclusion for any lakes but an outcome dictated by the interaction between wind speed, wind duration, basin morphometry, fetch distribution and sediment characteristics, among other factors (Sly, 1978; Evans, 1994). This, once more, brings to the fore the abovementioned debate over simplistic versus complex approaches. The finding emphasizes the point that while generalizations are convenient, it is necessary to keep in mind that there are situations where a holistic approach is more appropriate.

Climate change versus anthropogenic disturbance of the catchment
As introduced in Chapter 1, one of the problems occupying scientists working on Lake Victoria is the identification of the root causes for the decline of the lake's ecosystem. Connected to this is the question of quantifying how much of the present condition of the lake is the result of eutrophication, and how much is due to climate change. The results of the study provide evidence for the influence of both, but do not allow for differentiation or quantification of their contributions.

Chapters 4 and 6 provided evidence of eutrophication, a phenomenon identified also by others (Hecky, 1993; Hecky *et al.*, 1994; Ochumba, 1990; Bugenyi and Magumba, 1996; Verschuren *et al.*, 2002). I submit that eutrophication may have amplified and speeded up a change that was already underway, probably from climate variability. Comparing my findings with the work of others covering longer time spans has led me to conclude that in the medium- to long-term (centuries to millennia), climate change may be the main driver of ecosystem change, but in the short-term (decades), human activity may have a greater influence on ecosystem dynamics.

There is substantial evidence in the literature in support of such a view. Paleolimnological records spanning much longer periods than covered in this study (as early as ca. 17.5 ka) clearly show the present shift in signals (the $\delta^{13}C$ composition and C/N ratio) to be part of an earlier trend commencing in the Holocene period (Johnson *et al.*, 2000; Talbot and Lædal, 2000). According to a 10,000-year diatom fossil record examined by Stagger *et al.*, (1997), deep and continuous mixing of the lake waxed and waned repeatedly over periods of centuries. Other studies have shown that the paleoclimate of the region frequently shifted between extreme wet and dry conditions, with attendant fluctuations in lake levels (Yin and Nicholson, 1998; Nicholson, 1999; Verschuren *et al.*, 2000). A number of authors (Kendall, 1969; Talbot and Livingstone, 1989; Ngobi *et al.*, 1998; Johnson *et al.*, 2000; Talbort and Lædal, 2000) maintain that the lake underwent a series of major desiccation events in its history. Such dramatic climatic fluctuations can be expected to have a profound influence on the ecological dynamics of lakes within the region. Indeed, it has been argued (Johnson *et al.*, 2000) that some 9800 to 7500 years ago, from purely natural causes, limnological conditions in the lake deteriorated to a state comparable to that of today. Thus, it would seem that climate change plays a much larger role in regulating the condition of the lake than was previously assumed. Therefore, a closer examination of the role of climate change in the present condition of the lake should form the focus of future research in this area.

To say that climate change is important is not to downplay the impacts of recent human activities. My research has provided additional substantiation on the role of man in the deterioration of the lake. While climate change may not be entirely within the control of man, anthropogenic impacts are, and must be regulated rigorously.

Climate change and lake productivity

A pertinent question on the subject of climate change is its impact on the functioning of aquatic ecosystems. Scientists working on Lake Tanganyika have suggested that climate change diminishes the productivity of aquatic ecosystems (O'Reilly *et al.*, 2003; Verburg *et al.*, 2003). Is this the rule? Does climate change invariably lead to decreased productivity in all aquatic ecosystems? General conception (Goldman *et al.*, 1989; Carpenter *et al.*, 1992; Magnuson *et al.*, 1997), in disagreement with this presumption, maintains that contrasting outcomes may result from different systems, even if they are in close proximity, due to variability in modifying influences such as morphometry, position within the hydrological flow system, landuse in the watershed and assemblages of aquatic biota. The effects of global climate change operate through local weather parameters such as temperature, wind, rain, snow and water currents, as well as interactions among these. Climate variations, acting through the forcing factors, drive exchanges of heat momentum and water vapour, which ultimately determine growth, recruitment and migration patterns in ecosystems (Stenseth *et al.*, 2002).

Lakes Victoria and Tanganyika have many features in common. Both lie in the same geographical region (East Africa) and are controlled by the same regional environmental and climatic influences. In the two great lakes, climate change has produced a reduction in windiness, increase in mean annual water temperature, decrease in vertical mixing and increase in the duration of hypolimnetic anoxia (Hecky, 1993; Hecky *et al.*, 1994; O'Reilly *et al.*, 2002). Eutrophication has also been ongoing in the two lakes, as human populations in their catchments have grown (Hecky, 1993; Cohen *et al.*, 2005).

However, the two lakes have exhibited strikingly contrasting responses to climate change and eutrophication, thus providing a potent illustration of the above conception. While increasing nutrients inflows have produced an increase in primary productivity in Lake Victoria (Hecky, 1993, Lehman and Branstrator, 1993; Mugidde, 1993; Chapter 6, this study), they have produced a decrease in primary productivity in Lake Tanganyika (O'Reilly *et al.* 2003; Verburg *et al.*, 2003).

My results have led me to concur with general thinking, and to further expound on the submission by Verschuren (2003) that contrasting outcomes in the two lakes have arisen from differences in the frequency and spatial extent of nutrient upwelling and entrainment in surface waters, which I have attributed, in turn, to differences in morphometry. In tropical lakes, nutrient supply from the atmosphere, rock weathering or riverine sources cannot keep pace with the high rates of primary and secondary production due to year-round high temperatures (Verschuren, 2003). In such situations, mixing events are of crucial importance in recycling nutrients and maintaining lake productivity. In the African great lakes, mixing occurs during windy seasons from a combination of evaporative cooling of surface waters, thermocline tilting and mechanical agitation by wind (Spigel and Coulter, 1996).

I have argued (see Chapter 5) that in the relatively shallow Lake Victoria, sediment-bound nutrients originating in the catchment and settling on the lake floor are episodically returned to surface waters where they stimulate greater phytoplankton production. In the deeper and steep-sided Lake Tanganyika, seasonal mixing affects only the upper reaches of the deep-water nutrient reservoir, and most nutrients entering the lake are lost from the ecosystem after a single cycle through the food web. Thus, productivity in Lake Tanganyika is highly sensitive to changes in mixing efficiency. With decreased windiness and stronger stratification accompanying global climate change, the flux of sediment-bound nutrients to the epilimnion of Lake Tanganyika has considerably decreased, producing the observed decline in lake productivity.

Tropical lacustrine wetlands and atmospheric carbon sequestration

Anthropogenic CO_2 emissions from the burning of fossil fuels have upset the natural carbon biogeochemical cycle and brought about global warming (Stott *et al.*, 2000; Grace, 2004; Ehleringer *et al.*, 2005). Today ecologists and environmentalists are challenged to identify terrestrial and aquatic ecosystems that act as carbon reservoirs so that they can be managed in such a manner as to enhance their storage functions as a way of mitigating the impacts of global warming.

Are tropical freshwater systems of any importance in the temporary or permanent removal of carbon from the atmosphere? Previously, such a possibility was treated as remote or unlikely on account of the presumed rapid decomposition of organic matter within tropical lakes (Einsele *et al.*, 2001; Killops and Killops, 2005). However, this thesis has shown that some areas within Lake Victoria, a tropical lake straddling the equator, preserve sediments with a rich content of organic matter of terrestrial and aquatic origin. In my view, this underscores a potential for tropical lakes to contribute to the sequestration of atmospheric carbon and to amelioration of global warming. This vital function should serve as an added justification for their protection from wanton destruction.

The results presented in Chapters 5 and 6 show that organic matter preservation (inferred from the organic carbon content of bulk sedimentary organic matter) is highest adjacent to densely vegetated shorelines and within sheltered locations. This, it has been argued in the thesis, results from high biomass production by shoreline wetlands in the tropics, and by occurrence of shoreline vegetation in well-sheltered locations in which there is little wind and wave action and, therefore, minimal disturbance of buried organic matter. Thus, the interaction between productive terrestrial and littoral systems with lake morphometry and hydrodynamics may produce a set of unique conditions that enhance the function of lakes as carbon reservoirs.

What is the most critical intervention for Lake Victoria?

Now we come to the overriding issue put forward in Chapter 1: the most critical intervention for Lake Victoria. The foregoing discussion leads one to expect that this would be the reduction in nutrient inflows. It has been suggested above that nutrient inflows from the catchment upset the delicate natural balance in the lake's nearshore systems leading to the reduction in cover or disappearance of euhydrophytes, and with them fishes and other aquatic fauna. Sediment studies conducted as part of this work suggest the episodic occurrence of strong mixing events and internal currents with the potential to return sediment-bound nutrients to the water column and perpetuate nuisance algae growths. Thus, a reduction in nutrient inflows would seem like a good beginning point for lake restoration.

However, the reduction in nutrient loading, straight forward as it may seem, is a difficult undertaking requiring the implementation of a suite of cross-cutting actions such as control of soil erosion from hillsides and farmlands, control of stocking levels on communal rangelands, control of bush burning, reforestation of bare lands, latrine construction in coastal settlements, and treatment of municipal and industrial effluents (Sas and Ahlgren, 1989; Jeppesen *et al.*, 1999; Reynolds and O'Sullivan, 2005). Furthermore, considering that the phytoplankton-dominated state is a self-stabilising equilibrium, nutrient reductions will not immediately yield a return to a clear-water state (Sas and Ahlgren, 1989; Jeppesen *et al.*, 1991; Genkai-Kato and Carpenter, 2005). Thus, interventions must not only traverse sectors but also span years.

As well as introducing measures to reduce nutrient loading from the catchment, it is imperative to rehabilitate and protect the wetlands of the lake as a way of restoring their natural processes and functions and sustaining the productive fisheries of the lake.

References

Bengtsson, L., Hellström, T., and Rakoczi, L. (1990) Redistribution of sediments in three Swedish lakes. *Hydrobiologia*, 192: 167-181.

Bertness, M.D. (1999) *The Ecology of Atlantic Shorelines*, Sinauer Associates, Sunderland, Massachussetts 465 pp.

Blindow, I., Anderson, G., Hargeby, A., and Johansson, S. (1993) Long-term pattern of alternative stable states in 2 shallow eutrophic lakes. *Freshwater Biology*, 30(1): 159-167.

Bugenyi, F.W.B., and Magumba, K.M. (1996) The present physico-chemical ecology of Lake Victoria, Uganda. In: *The Limnology, Climatology and Paloeclimatology of the East African Lakes*, T.C. Johnson and E.O. Odada (eds). Gordon and Breach, Amsterdam pp. 141-154.

Carpenter, S.R., Fisher, S.G., Grimm, N.B., and Kitchell, J.F. (1992) Global change and freshwater ecosystems. *Annual review of ecology and systematics*, 23: 119-139 .

Clouet, J.F., and Fouque, J.P. (1994) Spreading of a pulse travelling in random media. *Annals Applied Probability*, 4: 1083-1097.

Cohen, A.S., Palacios-Fest, M.R., Msaky, E.S., Alin, S.R., McKee, B., O'Reilly, C.M., Dettman, D.L., Nkotagu, H., and Lezzar, K.E. (2005) Paleolimnological investigations of anthropogenic environmental change in Lake Tanganyika: IX. Summary of paleorecords of environmental change and catchment deforestation at Lake Tanganyika and impacts on the Lake Tanganyika ecosystem. *Journal of Paleolimnology*, 34: 125-145.

Denny, P. (1972) Sites of nutrient absorption in aquatic macrophytes. *Journal of Ecology*, 60: 819-829.

Dieter, C.D. (1990) The importance of emergent vegetation in reducing sediment resuspension in wetlands. *Journal of Freshwater Ecology*, 5: 467-474.

Ehleringer, J.R., Cerling, T.E., and Dearing, M.D.

(2005) *A History of Atmospheric CO₂ and Its Effects on Plants, Animals and Ecosystems*, Springer-Verlag, New York 482 pp.

Einsele, G., Yan, J., and Hinderer, M. (2001) Atmospheric carbon burial in modern lake basins and its significance for the global carbon budget. *Global and Planetary Change*, 30: 167-195.

Evans, R.D. (1994) Empirical evidence of the importance of sediment resuspension in lakes. *Hydrobiologia*, 284(1): 5-12.

Genkai-Kato, M., and Carpenter, S.R. (2005) Eutrophication due to phosphorous recycling in relation to lake morphometry, temperature and macrophytes. *Ecology*, 86(1): 210-219.

Goldman, C.R., Jassby, A., and Powell, T. (1989) Interannual fluctuations in primary production: meteorological forcing at two subalpine lakes. *Limnology and Oceanography*, 34(2): 310-323.

Grace, J. (2004) Presidential address: understanding and managing the global carbon cycle. *Journal of Ecology*, 92: 189-202.

Hecky, R.E., Bugenyi, F.W.B., Ochumba, P.O.B., Talling, J.F., Mugidde, R., Gophen, M., and Kaufman, L. (1994) Deoxygenation of the deep water of Lake Victoria, East Africa. *Limnology and Oceanography*, 39(6): 1476-1481.

Hecky, R.E. (1993) The eutrophication of Lake Victoria. *Verh. Internat. Verein. Limnol.*, 25: 39-48.

Holman, R.A., and Bowen, A.J. (1979) Edge waves on complex beach profiles. *Journal of Geophysical Research*, 84(C10): 6339-6346.

Hutchinson, G.E. (1975) *A Treatise on Limnology*, John Wiley and Sons, New York.

James, W.F., and Barko, J.W. (1990) Macrophyte influences on the zonation of sediment accretion and composition in a north-temperate reservoir. *Archiv fur Hydrobiologie*, 120: 129-142.

Jense, J.H. (1997) A model of nutrient dynamics in shallow lakes in relation to multiple stable states. *Hydrobiologia*, 342/343: 1-8.

Jeppesen, E., Kristensen, P., Jensen, J.P., Sondergaard, M., Mortensen, E., and Lauridsen, T. (1991) Recovery resilience following a reduction in external phosphorous loading of shallow, eutrophic Danish Lakes: duration, regulating factors and method of overcoming resilience. *Memorie dell'Istituto Italiano di Idrobiologia*, 48: 127-148.

Jeppesen, E., Jensen, J.P., Sondergaard, M., Lauridsen, T., Pedersen, L.J., and Jensen, L. (1997) Top-down control in freshwater lakes: the role of nutrient state, submerged macrophytes and water depth. *Hydrobiologia*, 342-343: 151-164 .

Jeppesen, E., Sondergaard, M., Kronvang, B., Jensen, J.P., Svendsen, L.M., and Lauridsen, T. (1999) Lake and catchment management

in Denmark. *Hydrobiologia*, 395/396: 419-432.

Johnson, T.C., Kelts, K., and Odada, E.O. (2000) The Holocene history of Lake Victoria. *Ambio*, 29(1): 2-11.

Keddy, P.A. (1985) Wave disturbance on lakeshores and the within-lake distribution of Ontario's Anlantic coastal plain fora. *Canadian Journal of Botany*, 63: 656-660.

Keddy, P.A. (2000) *Wetlands - Principles and Conservation*, Cambridge University Press, UK 614 pp pp.

Kendall, R.L. (1969) An ecological history of the Lake Victoria basin. *Ecological Monographs*, 39: 121-176.

Killops, S.D., and Killops, V.J. (2005) *Introduction to Organic Geochemistry*, Blackwell Publishing, Malden, USA 393 pp pp.

Knowlton, N. (1992) Thresholds and multiple stable states in coral reef community dynamics. *Integrative and comparative biology*, 32(6): 674-682.

Kufel, L., and Ozimek, T. (1994) Can *Chara* control phosphorous cycling in Lake Luknajno (Poland)? *Hydrobiologia*, 276: 277-283.

Lehman, J.T., and Brandstrator, D.K. (1993) Effects of nutrients and grazing on phytoplankton of Lake Victoria. *Verh. Internat. Verein. Limnol.*, 25: 850-855.

Leuttich, R.A., Harleman, Jr.D.R.F., and Somyódy, L. (1990) Dynamic behaviour of suspended sediment concentrations in a shallow lake perturbed by wind events. *Limnology and Oceanography*, 35(5): 1050-1067.

Lowe, E.F., Battoe, L.E., Coveney, M.F., Schelske, C.L., Havens, K.E., Marzolf, E.R., and Reddy, K.R. (2001) The resoration of Lake Apopka in relation to alternative stable states: an alternative view to that of Bachmann et. al. (1999). *Hydrobiologia*, 448(1-3): 11-18.

Maas, L.R.M., Benielli, D., Sommeria, J., and Lam, F.-P.A. (1997) Observation of an internal wave attractor in a confined stably stratified fluid. *Nature*, 388: 557-561.

Madsen, T.V., and Cedergreen, N. (2002) Sources of nutrients to rooted submerged macrophytes growing in a nutrient-rich stream. *Freshwater Biology*, 47: 283-291.

Magnuson, J.J., Webster, K.E., Assel, R.A., Bowser, C.J., Dillon, P.J., Eaton, J.G., Evans, H.E., Fee, E.J., Hall, R.I., Mortsch, L.R., Schindler, D.W., and Quinn, F.H. (1997) Potential effects of climate changes on aquatic systems: Laurential great lakes and Precambrian shield region. *Hydrological Processes*, 11(8): 825-871.

Mei, C.C. (1989) *The Applied Dynamics of Ocean Surface Waves*, World Scientific, Singapore.

Meyer, M.L., De Haan, M.W., Breukelaar, A., and Buitenveld, H. (1990) Effects of biomanipulation in shallow lakes: high transparency caused by zooplankton, macrophytes or lack of benthivorous fish? *Hydrobiologia*, 200/201: 303-317.

Mothes, G. (1987) The influence of a submerged macrophyte mat on the pelagic balance of substances in the lake. *Acta Hydrochimica et Hydrobiologica*, 15(2): 161-166.

Mugidde, R. (1993) The increase in phytoplankton primary productivity and biomass in Lake Victoria (Uganda). *Verh. Internat. Verein. Limnol.*, 25: 846-849.

Nachbin, A., and Solna, K. (2003) Apparent diffusion due to topographic microstructure in shallow waves. *Physics of Fluids*, 15(1): 66-77.

Nakai, S., Inoue, Y., Hosomi, M., and Murakami, A. (1999) Growth inhibition of blue-green algae by allelopathic effects of macrophytes. *Water Science and Technology*, 39(8): 47-53.

Nakai, S., Inoue, Y., Hosomi, M., and Murakami, A. (2000) *Myriophyllum spicatum*-relased allelopathic polyphenols inhibiting growth of blue-green algae *Microcystis aeruginosa*. *Water REsearch*, 34(11): 3026-3034.

Ngobi, G.N., Kelts, K., Johnson, T.C., and Solheid, P.A. (1998) Environmental magnetism of the late Pleistocene-Holocene sequences from Lake Victoria, East Africa. In: *Environmental Change and Response in East African Lakes*, J.T. Lehman (ed). Kluwer Academic Publishers, Dordrecht pp. 59-74.

Nicholson, S.E. (1999) Historical and modern fluctuation of Lakes Tanganyika and Rukwa and their relationship to rainfall variability. *Climate Change*, 41: 53-71.

Nielsen, P. (1992) *Coastal Bottom Boundary Layers and Sediment Transport*, World Scientific, New Jersey 324 pp.

Norlin, I.J., Bayley, S., and Ross, L.C.M. (2005) Submerged macrophytes, zooplankton and the predominance of low- over high-chlorophyll states in western boreal, shallow-water wetlands. *Freshwater Biology*, 50: 868-881.

O'Reilly, C.M., Hecky, R.E., Cohen, A.S., and Plisnier, P.-D. (2002) Interpreting stable isotopes in food webs: Recogonizing the role of time averaging at different trophic levels. *Limnology and Oceanography*, 47(1): 306-309.

O'Reilly, C.M., Alin, S.R., Plisnier, P.-D., Cohen, A.S., and McKee, B.A. (2003) Climate change decreases aquatic ecosystem productivity of Lake Tanganyika, Africa. *Nature*, 424: 766-768.

Ochumba, P.B.O. (1990) Massive fish kills in the Nyanza gulf of Lake Victoria, Kenya. *Hydrobiologia*, 208: 93-99.

Petrevu, U., and Oltman-Shay, J. (1998) Influence functions of large wave propagation over a nonpolar bottom bathymetry. *Physics of Fluids*, 10(1): 330-332.

Petticrew, E.L., and Kalff, J. (1992) Water flow and clay retention in submerged macrophyte beds. *Canadian Journal of Fisheries and Aquatic Sciences*, 49: 2483-2489.

Reading, H.G., and Levell, B.K. (1996) Controls on the sedimentary record. In: *Sedimentary Environments: Processes, Facies and Stratigraphy*, H.G. Reading (ed). Blackwell Science, Oxford pp. 5-25.

Reynolds, C.S., and O' Sullivan, P. (2005) *The Lakes Handbook: Lake Restoration and Rehabilitation*, Blackwell Science, Oxford 568 pp pp.

Rowan, D.J., Kalff, J., and Rasmussen, J.B. (1992) Estimating the mud deposition boundary depth in lakes from wave theory. *Canadian Journal of Fisheries and Aquatic Sciences*, 49: 2490-2497 .

Sas, H., and Ahlgren, I. (1989) *Lake Restoration by Reduction of Nutrient Loading: Expectations, Experiences, Extrapolations*, Academia Verlag Richarz, St. Augustine 497 pp pp.

Scheffer, M. (1990) Multiplicity of stable states in freshwater systems. *Hydrobiologia*, 200/201: 475-486 .

Scheffer, M. (1998) *Ecology of Shallow Lakes*, Chapman & Hall, London 357 pp.

Scheffer, M. (1999) The effect of aquatic vegetation on turbidity: how important are the filter feeders? *Hydrobiologia*, 408/409: 307-3163.

Scheffer, M., and Jeppesen, E. (1998) Alternative stable states. In: *Structuring Role of Submerged Macrophytes in Lakes*, E. Jeppesen, M. Sondergaard, M. Sondergard and K. Kristoffersen (eds) . Springer-Verlag, New York pp. 397-406.

Sly, P.G. (1978) Sedimentary processes in lakes. In: *Lakes: Chemistry, Geology Physics*, A. Lerman (ed). Springer Verlag, New York pp. 65-89.

Spence, D.H.N. (1982) The zonation of plants in freshwater lakes. *Advances in Ecological Research*, 12: 37-125.

Spigel, R.H., and Coulter, G.W. (1996) Comparison of hydrology and physical limnology of the East African Great Lakes: Tanganyika, Malawi, Victoria, Kivu and Turkana (with reference to some North American Great Lakes). In: *The Limnology, Climatology and Paloeclimatology of the East African Lakes*, T.C. Johnson and E.O. Odada (eds). Gordon and Breach Publishers, Amsterdam pp. 103-140.

Stager, J.C., Cumming, B., and Meecher, L. (1997) A high-resolution 11,400-year diatom record from Lake Victoria, East Africa. *Quaternary Research* , 47: 81-89.

Stenseth, N.C., Mysterud, A., Ottersen, G., Hurrell, J.W., Chan, K., and Lima, M. (2002) Ecological effects of climate fluctuations. *Science*, 297(5585): 1292-1296.

Stott, P.A., Tett, S.F.B., Jones, G.S., Allen, M.R.,

Mitchell, J.F.B., and Jenkin, G.F. (2000) External control of 20th century temperature by natural and anthropogenic forcings. *Science*, 290: 2133-2137.

Talbot, M.R., and Livingstone, D.A. (1989) Hydrogen index and carbon isotopes of lacustrine organic matter as lake level indicators. *Palaeogeography, Palaeoclimatology, Palaeoecology*, 70((1-3)): 121-137.

Talbot, M.R., and Lærdal, T. (2000) The Late Pleistocene-Holocene paleolimnology of Lake Victoria, East Africa, based on elemental and isotopic analyses of sedimentary organic matter. *Journal of Paleolimnology*, 23: 141-164.

Thurman, H.V., and Trujillo, A.P. (2004) *Introductory Oceanography*, Prentice Hall, New Jersey 608 pp.

Timms, R.M., and Moss, B. (1984) Prevention of growth of potentially dense phytoplankton populations by zooplankton grazing in the presence of zooplanktivorous fish in a shallow wetland ecosystem. *Limnology and Oceanography*, 29: 472-486.

Van de Koppel, J., Herman, P.M.J., Thoolen, P., and Heip, C.H.R. (2001) Do alternate stable states occur in natural ecosystems? Evidence from a tidal flat. *Ecology*, 82: 3449-3461.

Van den Berg, M.S., Coops, H., Meijer, M.L., Scheffer, M., and Simons, J. (1998) Clear water associated with a dense *Chara* vegetation in the shallow and turbid Lake Veluwe, the Netherlands. In: *Structuring Role of Submerged Macrophytes in Lakes*, E. Jeppesen, M. Sondergaard, M. Sondergard and K. Kristoffersen (eds). Springer-Verlag, New York pp. 339-352.

Van Donk, E., Gulati, R.D., Iedama, A., and Meulemans, J.T. (1983) Macrophyte-related shifts in the nitrogen and phosphorous contents of the different trophic levels in a biomanipulated shallow lake. *Hydrobiologia*, 251: 19-26.

Van Donk, E., and Van de Bund, W.J. (2002) Impact of submerged macrophytes including charophytes on phyto-and zooplankton communities: allelopathy versus other mechanisms. *Aquatic Botany*, 72(3-4): 261-274.

Verburg, P., Hecky, R.E., and Kling, H. (2003) Ecological consequences of a century of warming in Lake Tanganyika. *Science*, 301(5632): 505-507.

Verschuren, D. (2003) The heat on Lake Tanganyika. *Nature*, 424: 731-732.

Verschuren, D., Johnson, T.C., Kling, H.J., Edgington, D.N., Leavit, P.R., Brown, E.T., Talbot, M.R., and Hecky, R.E: (2002) History and timing of human impact on Lake Victoria, East Africa. *Proceedings of the Royal Society of London*, 269: 289-294.

Verschuren, D., Laird, K., and Cumming, B.F. (2000) Rainfall and drought in equatorial East Africa during the past 1,100 years. *Nature*, 403: 410-414.

Wium-Andersen, S. (1987) Allelopathy among aquatic plants. *Archiv fur Hydrobiologie*, 27: 167-172.

Wium-Andersen, S., Anthoni, U., Christophersen, C., and Houen, G. (1982) Allopathic effects of phytoplankton by substances isolated from aquatic macrophytes (Charales). *Oikos*, 39: 189-190.

Yin, X., and Nicholson, S.E. (1998) The water balance of Lake Victoria. *Hydrological Sciences Journal*, 43(5): 789-811.

Chapter 8

Conclusions and recommendations

Conclusions and recommendations

Conclusions

In this study the way in which shoreline wetlands and sediments of northern Lake Victoria are spatially distributed, factors regulating their distribution and possible ways in which they influence the ecological dynamics of the lake were examined. The aforementioned are among the many areas considered not well understood about the functioning of the lake. The underlying goal of the study was to build upon what was already known so that with the increased mass of knowledge, management interventions needed to halt the lake's decline could more easily be identified. Below are the conclusions of the study.

Spatial distribution of shoreline wetlands and factors causing patterning
1. Wave exposure, I have noted, is an important factor regulating the spatial distribution of shoreline vegetation in Lake Victoria. As a consequence of its influence, shoreline vegetation in northern Lake Victoria mostly occupies stretches of shoreline shielded from wind and wave action by coastal islands or sites well hidden from wind by convolutions of the lake margin. I have shown that the area of shoreline swamps declines exponentially with increasing wave exposure.

2. The bay area regulates also the spatial distribution of shoreline vegetation. As the area of bay increases, the lakeward front of shoreline vegetation moves further and further back from the bay mouth. Bay area in combination with wave exposure explained 64.4% of the variance in the lakeward limit of vegetation in the study sites of this research.

3. By assessing conditions within habitats occupied by floating mats and relating those conditions to general ecological theory on the factors controlling the distribution of littoral vegetation, I came to the conclusion that among the factors critical for development of floating vegetation mats in the shoreline swamps of Lake Victoria are (a) shelter from physical disturbance, particularly wind-wave action; and (b) stability in the hydrological regime, particularly absence of large and rapid water level fluctuations.

4. Lake Victoria has had a reasonably stable hydrological regime in the past 100 years. I contend that the hydrological stability could be part of the reason Lake Victoria has large areas of floating emergent swamps.

5. Of the many potential mechanisms outlined in literature by which floating mats of emergent vegetation are initiated, the likely ones for Lake Victoria, which I deduced from assessment of conditions in the shoreline areas are (a) lake-ward expansion of emergent macrophytes as a stage in succession; and (b) detachment of a light, and probably gas-filled, mat of emergent vegetation from the lake floor by gradually rising lake levels.

6. Shoreline emergent vegetation in the study area growing in the floating-mat form has a depth range (0 – 6.5 m) that is three times greater than the depth range of emergent vegetation rooted on the bottom (0 – 2 m).

7. My research has shown that emergent vegetation with floating mats mostly occur in localities where conditions favourable for floating mat establishment (enumerated in 3 above) are present. In such locations, emergent vegetation with floating mats is dominant over bottom-rooted emergents. In other locations, bottom-rooted emergents are dominant over floating-mat emergents.

8. The net influence of the mat, from the results of this research, seems to be in increasing vertical (cross-shore) distribution and reducing horizontal (alongshore) distribution of mat-forming emergent vegetation. I have argued that, other factors remaining constant, whether the influence of the mat will lead to a larger or smaller cover of mat-forming emergents as compared to bottom-rooted emergents will depend on the relative proportion of hydrologically stable, low-energy locales to hydrologically unstable, high-energy locales along a given stretch of shore.

Fish species decline: possible connection to littoral wetland decline

9. I believe there is a strong possibility that fish species declined in Lake Victoria partially as a result of the degradation of euhydrophyte-dominated littoral habitats following the eutrophication of the lake. The decline in euhydrophytes caused a loss of habitat, food and refuge for fish.

10. A second mechanism by which eutrophication could have caused the decline in fish species, which I have highlighted in this research, is the development of chemically unfavourable conditions in shoreline wetlands which force fish to migrate to open waters where they are at greater risk of predation.

11. This study has shown that dissolved organic substances, especially natural yellow substances (humic and fulvic acids), are present at high levels in the nearshore waters of Lake Victoria. I have suggested that their increase from relatively low levels in the 1960s to the present high levels could have contributed to the decline in euhydrophyte cover in the lake.

Spatial distribution of surficial sediments and mechanisms of sediment transport

12. The surficial sediments of Lake Victoria range from very coarse sands to fine clays. In agreement with present generalisations, I found that most fine-grained sediments occur in the deep central parts of the basin while coarse-grained sediments occur along basin rims. However, there are regions where the sediment distribution pattern is not in accord with these generalisations, notably on the northern coast where fine-grained sediments fringe the shore, and the western part of the basin, where a belt of coarse-grained sediments extends from the margins to far offshore regions.

13. The distribution pattern of surficial sediment has led me to conclude that wind induced wave-resuspension is the dominant sediment resuspension mechanism in Lake Victoria, but apparently not the only important means by which sediment is resuspended and transported. Alongshore and cross-shore currents

generated during extreme storms appear to play a significant role in the transport of sediments in the lake. On the basis of these findings, I submit that in large shallow lakes located in regions with strong prevailing winds, current-resuspension mechanisms, and combined wave- and current-resuspension mechanisms should be considered alongside wave-resuspension mechanisms.

14. The results of the prediction of the location of sedimentological zones in Lake Victoria for different wind conditions suggest that in large shallow lakes, a centrally focused pattern of sediment distribution only results when prevailing winds are of variable direction during the episodic events that resuspend and move sediments. When prevailing winds are unidirectional, a linear pattern of sediment distribution results, with fine-grained sediments accumulating at the upwind end of the axis and coarse-grained sediments accumulating at the downwind end of the axis.

15. A systematic progression of sediment characteristics along depth and fetch gradients in lakes only occurs where the sorting effects of wave-resuspension have not been cancelled, confounded or overwritten by other sediment resuspension mechanisms.

16. In my view, the basin-scale mechanisms that distribute sediments in Lake Victoria are: (a) wind-driven epilimnetic mixing and lake-wide deposition of a homogenous suspension; and (b) current-mediated alongshore and cross-isobath movement of sediments.

17. The data examined in this study shows that the wind regime over Lake Victoria is characterised by co-dominance of the Northeastern and Southeastern components of the easterlies from the Indian Ocean.

18. The finding of an orderly progression in surficial sediment characteristics along a linear belt starting from the mouth of River Kagera has led me to conclude that during certain periods of the year, the Kagera River, probably helped by storm-generated currents, crosses the breadth of Lake Victoria to exit through the Nile at Jinja.

Causal factors of ecological change in Lake Victoria

19. Current theories that cite eutrophication and climate change as the leading causes for the ecosystem changes in Lake Victoria are partially accurate and partially flawed.

20. In agreement with current theories, the limnological record examined in this research shows eutrophication to have occurred and produced an increase in phytoplankton productivity starting in the first quarter of the previous century. Furthermore, denitrification appears to have occurred, and produced isotopically heavy N_2. However, different parts of the lake responded differently to the stimulus of increased nutrient inflows indicating regional differences in initial states and buffering capacities. The data shows a discernible movement of impacts from nearshore to offshore areas, with time lags of a few decades in the manifestation of impacts at the different sites.

21. The amount of N-fixing cyanobacteria material in the most recent sediments of Lake Victoria is not greater than that of non-N-fixing phytoplankton.

22. It is my belief that eutrophication has not initiated a new shift in lake conditions, but amplified and accelerated a trend that was already underway due to more natural causes, such as a gradual shift in global climatic conditions.

23. Contrary to previous interpretations, I contend that the sedimentary organic matter of Lake Victoria is not entirely comprised of algal matter but has varying proportions of terrestrial, swamp and algal plant material.

Recommendations

Following from the above findings and preceding general discussion on their significance (Chapter 7), a number of recommendations to address the problem of the lake's decline have been made. These are presented below and are of two categories: management and further research.

Management of Lake Victoria
1. The most urgent intervention for Lake Victoria is the reduction of pollution loads from the catchment. It is recommended for actions in this area to be expedited.

2. Measures should immediately be put in place to restore degraded wetlands and prevent destruction of new wetlands so as to sustain their important functions.

3. To the extent possible, fluctuations in the level of Lake Victoria should be kept within the natural range of amplitudes recorded before significant human interference. Management of the lake's levels by man that results in large and sudden draw downs will affect the community of shoreline plants, with dire consequences for aquatic fauna including commercially important fish, and the livelihoods of lakeside human communities.

4. It is recommended to commence systematic and periodic monitoring of nearshore waters and associated wetlands. Parameters to be monitored should include vegetation cover and species composition, physico-chemical and optical conditions of nearshore waters, and abundance and species composition of coastal fauna.

Further research
The findings of this study are only a small fraction of the total knowledge needed to manage the Lake Victoria ecosystem. Further investigation must therefore be vigorously pursued. Below are recommended areas for further research.
1. A combined hydrodynamics and sediments study to refine the basin-scale sediment transport mechanisms identified in this study. The sediment study should cover the entire lake and use a finer sampling grid than used in this study (perhaps a 10 x 10 km grid). For the purpose of this study and long-term monitoring of the lake, the network of meteorological stations on the lake should be increased.

2. One of the objectives of the Lake Victoria Environmental Management Project is to develop a physical processes and water quality model for Lake Victoria. My results suggest that the lake occasionally approaches ideally-stirred behaviour, and lakewide recharge of nutrients from bottom sediments occurs. For the model to give realistic results, it is recommended that it incorporates such extreme hydrodynamic events.

3. Further research to quantify and distinguish between the contributions of the two main causes – eutrophication and climate change – to the present condition of Lake Victoria is recommended.

4. The results of this study suggest that the magnitude as well as the frequency of occurrence of wave disturbance events is important in regulating the shoreline distribution of vegetation. More research is recommended to examine the quantitative contribution and relative importance of magnitude and frequency of wave disturbance on shoreline vegetation distribution.

5. My research has indicated a potential for using wave exposure to predict absence/presence of shoreline plants. This could be a useful tool in lake restoration programmes. Further research is recommended to generate a database of empirical data that could be used to build predictive models.

6. My results suggest succession and detachment to be the initiation mechanisms for the floating mats of Lake Victoria, and that possession of a dual growth form confers a competitive advantage to plants. Further research into these areas is recommended.

Samenvatting

Dit proefschrift presenteert de resultaten van onderzoek aan de kustmoerassen en het sediment van de noordelijke regio van het Victoriameer. Dit meer, wat gedeeld wordt door de landen Kenia, Tanzania en Oeganda, is het grootste meer in Afrika en het op één na grootste meer ter wereld. Het herbergt een grote diversiteit aan aquatische planten en dieren en speelt een cruciale rol in de socio-economische ontwikkeling van de Oost- Afrikaanse regio. Het voorziet de bevolking in de regio onder andere van drinkwater, waterkrachtenergie, en goedkope dierlijke proteïnes.

In de afgelopen decennia is de toestand van het meer aanmerkelijk verslechterd. Veel van de vitale functies van het meer kunnen verloren raken als er geen maatregelen genomen worden om deze achteruitgang aan te pakken. Het is echter moeilijk de meest belangrijke ingrepen te definiëren, omdat er nog belangrijke gaten zijn in de kennis over het meer. De kustmoerassen en sedimenten behoren tot de gebieden waarover de kennis het meest beperkt is. Studies in andere meren geven aan dat deze moerassen en sedimenten mogelijk een belangrijke invloed hebben op de dynamiek van het ecosysteem in het meer.

Het doel van deze studie was te bepalen hoe kustmoerassen en sedimenten verdeelt zijn over het noordelijke gedeelte van het Victoriameer, welke factoren deze verdeling bepalen, en op welke manier ze het functioneren van het meer beïnvloeden. De studie beoogt enkele van deze kennisgaten op te vullen, om het bepalen van de juiste maatregelen voor het herstel van het meer te vereenvoudigen.

Het onderzoek toonde aan dat kustmoerassen zich voornamelijk bevinden in gebieden met lage wind- en golfexpositie. Naast golfexpositie is de grootte van de baaien een belangrijke regulerende factor voor de ruimtelijke distributie van kustvegetatie. Het Victoriameer bleek een stabiel hydrologisch regime en vele beschutte baaien te hebben langs zijn zeer onregelmatige kustlijn. Deze kenmerken leken verantwoordelijk te zijn voor de overvloed van kustmoerassen aan de randen van het meer. De resultaten suggereerden dat matvormende vegetatie een toename in verticale distributie (van kant naar water) en afname in horizontale distributie (langs de kust) van uit het water opkomende planten veroorzaakt.

Door een beoordeling van de condities in het meer in geselecteerde kustgebieden, bleken gebieden dicht bij de kustmoerassen een lage concentratie van opgeloste zuurstof, een hoge concentratie van opgelost organisch materiaal, hoge lichtuitdoving en een dagelijkse cyclus in ammoniak productie te hebben. De studie stelt dat de bovenstaande condities, die uitgesproken verschillend zijn van de condities in 1960, het gevolg zijn van eutrofiëring en geleid hebben tot degradatie van waterplant-gedomineerde litorale habitats. Deze litorale habitats, voornamelijk gebieden met een overvloedige bedekking van waterplanten, voorzien vis van een veilig toevluchtsoord, van voedsel en van geschikte broedlocaties. De vermoedelijke vernietiging van dit habitat kan, zoals wordt beargumenteerd in deze studie, een belangrijke bijdrage geleverd hebben aan de daling van vispopulaties in het Victoriameer.

De oppervlaktesedimenten van het Victoriameer, zo vond het onderzoek, zijn verdeeld op een ongewone manier. Hoewel de meeste fijnkorrelige sedimenten gevonden werden in diepe gebieden in het midden van het meer, en grofkorrelige sedimenten vooral voorkomen in ondiepe kustgebieden, waren er toch aanzienlijke gebieden waar geen overeenstemming gevonden werd met deze meer algemene verdeling. We vonden sterke aanwijzingen dat de sedimentverdeling mede bepaald werd door de oriëntatie van de heersende winden en stromingen. Opmerkelijk was een lineaire band van systematische gradatie in het karakter van het sediment, lopend van de westelijke naar de noordoostelijke oevers van het meer. Deze studie stelt, gebaseerd op de afwijkingen in sediment distributie, dat incidenteel sterke turbulente vermenging voorkomt in het meer, en dat waterstromen gegenereerd door stormen gereflecteerd worden naar het centrum van het meer door de aanwezigheid van de Sese eilanden. De studie concludeert dat sommige van de concepten van distributieverdeling, gebaseerd op kleine meren, niet goed van toepassing zijn op grote meren.

Om meer licht te werpen op de mogelijk oorzaken van de achteruitgang van het Victoriameer, onderzocht deze studie sedimentkernen uit het meer op enkele paleolimnologische signalen. De resultaten bevestigden vroegere studies die laten zien dat eutrofiëring de oorzaak is van de veranderingen, maar laat ook zien dat klimaatverandering een belangrijke bijdrage heeft geleverd. De resultaten van de studie suggereren dat het organische materiaal in het sediment niet uitsluitend van algen afkomstig kan zijn, maar een mengsel is van terrestrisch materiaal en algen. Geen aanwijzingen werden gevonden, waar dan ook, voor een verandering in dominantie tussen fytoplanktongroepen, en er was geen synchronisatie in de snelheid en afloop van de veranderingen in het stabiele isotoopsignaal. Het bovenstaande komt overeen met bestaande ideeën over de veranderingen in het Victoriameer. We beargumenteren dat de eutrofiëring veranderingen heeft versterkt die al begonnen waren als gevolg van klimaatveranderingen.

Met deze verbeterde kennis over het meer kunnen een aantal aanbevelingen gedaan worden met betrekking tot de achteruitgang van het Victoriameer. Deze studie laat zien dat een reductie van de nutriënten toevoer de meest belangrijke beheersmaatregel is, en maatregelen in deze richting dienen met zo min mogelijk vertraging genomen te worden. De studie heeft verder onderzoek gedaan op een aantal belangrijke velden, in het bijzonder het hydrodynamische regime van het meer, de mechanismen van sedimentdistributie, en het bepalen van de rol van klimaatsverandering in de achteruitgang van het ecosysteem van het Victoriameer.

About the Promovendus

Nicholas Azza was born in October 1966 in the tiny northern Uganda town of Metu in Moyo district. After receiving primary and secondary education, Azza joined Makerere University in Kampala in 1986 for a 3-year undergraduate course in Chemistry and Biochemistry. On graduation, he joined the Ministry of Water and Environment as a Water Analyst and was posted to his current duty station in Entebbe.

Between 1994 and 1996, Azza participated in the International Course in Water Quality Management of UNESCO-IHE Institute for Water Education obtaining first, a postgraduate diploma and then, a Master of Science degree in Water Quality Management, both with distinctions. He returned to work in Uganda before enrolling a few years later for the PhD Programme of UNESCO-IHE Institute for Water Education.

Azza still works for the Ministry of Water and Environment in Uganda and has 15 years of experience in water resources management and water pollution control in Uganda. He rose over the years to the rank of Principal Analyst and headed the Water Quality and Pollution Control Laboratory of the Ministry from 1993 till 1999 when he took leave to pursue further studies. During the time he headed the national laboratory, he designed and operationalised a national water quality monitoring network, directed the refurbishment, re-equipment and modernisation of the laboratory, oversaw the introduction of a quality control system conforming to international standards and supervised the analysis of over 5,000 water and waste water samples. Azza is a gazetted Environmental Inspector and a member of the National Technical Committee that advises the National Environment Management Authority (NEMA) of Uganda on matters related to environmental impact assessments (EIAs).

Between August 1997 and April 2000, Azza served as Deputy Project Coordinator of the Water Resources Assessment Project (WRAP). This was the start-up program for a long-term government plan to strengthen institutional capacity for water resources assessment and monitoring in Uganda.

Azza headed the sub-component called Management of Eutrophication of the World Bank/GEF-funded Lake Victoria Environmental Management Project (LVEMP) between 1996 and 1999. Through this programme, he became actively involved in the restoration efforts on Lake Victoria. Recent assignments to Azza in the Nile Basin (where Lake Victoria lies) include serving as National Liaison Officer for Uganda for the Kagera and Sio-Malaba-Malakisi integrated river basin management projects under the Nile Basin Initiative (NBI).

Training under the STEP Programme
(Sandwich-model PhD)

PhD Seminars
- 2002 (attended, chaired one session, made a presentation)
- 2003 (attended)
- 2005 (attended)

Topics and activities covered in the PhD seminars
- Presentation skills
- Scientific writing
- Dissertation writing
- Management of PhD research projects
- Use of citation software (Procite and Endnote)
- Role play (transboundary water resources management)
- Study tours (water and waste water works, hydraulic works, chemical management, port management, practitioners/consultants in integrated water resources management)

International Conferences and Conventions

2000   Attended the Second World Water Forum, held from 17-22 March 2000 in The Hague, The Netherlands.

1999   Attended the conference *Greening the Waters* held at Merseyside Maritime Museum, Liverpool, UK (November 1999).

Regional Conferences

2002   Participated in the First Regional Scientific Conference on Lake Victoria held in December 2002 in Kisumu, Kenya. Presented paper titled *Diel cycles of physico-chemical conditions in Lake Victoria.*

2001   Participated in the National Scientific Conference on Lake Victoria organised by the LVEMP in August 2001 in Mukono, Uganda. Presented two papers titled *Hydrodynamics of Lake Victoria*, and *Sediment characteristics and distribution in Northern Lake Victoria.*

2000   Participated in the scientific conference titled *Lake Victoria 2000: A New Beginning* organized by the Lake Victoria Fisheries Organization (LVFO) in Jinja, Uganda (June 2000)

Visiting Scientist Programs

1999   Visited the UK (November 1999) and held meetings/tutorials with:
- Dr. Nick Haycock – Quest Environmental, UK
- Prof. Brian Moss – School of Biological Sciences, University of Liverpool

- Dr. Jack Talling – Institute of Freshwater Ecology (IFE), Windermere Laboratory
- Prof. Colin S. Reynolds – IFE, Windermere Laboratory
- Dr. Ed Tipping – IFE, Windermere Laboratory

2002   Visited the Netherlands Centre for Estuarine and Marine Ecology (CEMO), Yerseke in March 2002 and:
- Toured CEMO research facilities
- Gave lunch-time presentation title *Eutrophication of Lake Victoria*
- Had a tutorial with Dr. Jack Middelburg and Dr. Peter Herman on organic geochemistry and sediment chronology

2002   Visited the Department of Chemical and Biosystem Sciences, University of Siena, Italy (April 2002) and:
- Met with Prof. Claudio Rossi
- Toured laboratory facilities of the Department
- Had a tutorial with Stefania Mazzuoli on laboratory analysis of humic acids using UV/Visible Scanning Spectroradiometer
- Had a tutorial with Dr. Steven Lioselle and Luca Bracchini on light attenuation measurements with the PUV-Profiling Ultraviolet Radiometer.

IHE Lunch Seminars
- Attended several lunch seminars in 1999, 2000 and 2005
- Made presentations in lunch seminars in March 2000, May 2005 and August 2005

Contribution to the Education Programme of UNESCO-IHE

2005   Gave a one-period lecture on *Natural wetlands for water treatment: case study Lake Victoria wetlands* to Msc students at IHE.

2004   Presented a video conference discussion paper titled *Natural wetlands for water treatment: case study Lake Victoria wetlands* from the Global Distance Leaning Centre in Kampala to an audience in UNESCO-IHE participating in the Wetlands for Water Quality module.

2003   Discussant in the video conference on the theme *"The Relevance of Human Resources/Capacity Building Needs for Public-Private Partnership (PPPs) in the Water and Sanitation Sector"* organized by The World Bank and UNESCO-IHE (October 2003) involving the GDLCs of IHE, Cairo-Egypt, Kampala-Uganda, Beijing-China, Manila-Phillipines, New Delhi-India, Colombo-Sri Lanka and Birzeit University –Palestine.

Printed and bound by CPI Group (UK) Ltd, Croydon, CR0 4YY

22/10/2024

01777642-0009